ALTERNATIVE WASTEWATER TREATMENT

LOW-COST SMALL SYSTEMS,
RESEARCH AND DEVELOPMENT

WATER SCIENCE AND TECHNOLOGY LIBRARY

ALTERNATIVE WASTEWATER TREATMENT. Low-Cost Small Systems, Research and Development.
Edited by A. S. Eikum and R. W. Seabloom ISBN 90-277-1430-4

ALTERNATIVE WASTEWATER TREATMENT

LOW-COST SMALL SYSTEMS, RESEARCH AND DEVELOPMENT

Proceedings of the Conference held at Oslo, Norway, September 7-10, 1981

Edited by

ARILD SCHANKE EIKUM
NIVA, Oslo

and

ROBERT W. SEABLOOM
University of Washington, Seattle

Sponsored by:
University of Washington, Seattle, Washington, U.S.A.
US Environmental Protection Agency (EPA), Cincinnati, Ohio, U.S.A.
Agricultural Research Council of Norway (NLVF), Ås, Norway
Norwegian State Pollution Control Authority (SFT), Oslo, Norway
Norwegian Institute for Water Research (NIVA), Oslo, Norway

D. REIDEL PUBLISHING COMPANY
Dordrecht, Holland / Boston, U.S.A. / London, England

Library of Congress Cataloging in Publication Data

CIP

DATA APPEAR ON SEPARATE CARD.

ISBN 90-277-1430-4

Published by D. Reidel Publishing Company,
P.O. Box 17, 3300 AA Dordrecht, Holland.

Sold and distributed in the U.S.A. and Canada
by Kluwer Boston Inc.,
190 Old Derby Street, Hingham, MA 02043, U.S.A.

In all other countries, sold and distributed
by Kluwer Academic Publishers Group,
P.O. Box 322, 3300 AH Dordrecht, Holland.

D. Reidel Publishing Company is a member of the Kluwer Group.

Printed in The Netherlands

CONTENTS

Program Committee:

R.W. Seabloom	University of Washington Department of Civil Engineering Environmental Engineering & Science, FX-10 Seattle, Washington 98195, USA
G.F. Kreissl	US Environmental Protection Agency (EPA) Municipal Environmental Research Laboratory Cincinnati, Ohio, 45268, USA
L. Bræin	Norwegian State Pollution Control Authority (SFT) P.O.Box 8100 Oslo-dep, Oslo 1, Norway
T. Østeraas	Agricultural Research Council of Norway (NLVF) 1432 Ås-NLH, Norway
A.S. Eikum	Norwegian Institute for Water Research (NIVA) P.O.Box 333, Blindern, Oslo 3, Norway

PREFACE

Following the end of World War II there was a major migration of population in the United States and Scandinavian countries to urban areas. As a result of this migration and in part due to the public works moratoria imposed during the war, a major program of sewer construction was instigated, which resulted in the collection and subsequent concentration of large volumes of wastewater at single discharge points. As the assimilative capacity of these receiving waters was exceeded, it led to or aggravated existing water pollution problems in these waters. To mitigate this degradation of water quality a massive program to construct wastewater treatment facilities was instigated. In addition, large amounts of money were spent on research to improve the technology of the conventional collection and treatment concept. In contrast, the wastewater disposal problem of the rural home owner received little attention, and in most cases the septic tank soil absorption system (ST-SAS) was the interim solution.

In recent years there has been a fundamental change in the population growth pattern in the US and Scandinavian countries. It appears that a great many people are moving back to rural areas where they seem to prefer the suburban or small town environment, yet at the same time want all the conveniences of urban life. The provision of proper wastewater disposal facilities presents a very perplexing problem, because the capital and operating costs of conventional sewers are usually financially impractical for rural areas.

A. S. Eikum and R. W. Seabloom (eds.), Alternative Wastewater Treatment, ix–x.

The sponsors of this conference;
University of Washington (Seattle, Washington, USA);
US Environmental Protection Agency (Cincinnati, Ohio, USA);
Agricultural Research Council of Norway; Norwegian State Pollution Control Authority; and Norwegian Institute for Water Research, have long recognized the need for alternative low cost small wastewater treatment systems. In the US and Norway about 30 percent of the population rely on some form of on-site wastewater disposal systems.

When properly designed, constructed, and maintained on a suitable site, the ST-SAS is an excellent method of disposing of liquid wastes. Unfortunately, conditions that limit the suitability of these systems, such as soil with low permeability, high ground water, shallow soils, and excessive slopes, are quite prevalent in Scandinavia and in certain parts of the US.

It was the purpose of this conference to present the latest information on and exchange research results on the design, construction, operation, and management of small low cost alternative wastewater treatment systems.

The speakers were internationally recognized experts with extensive experience in the small scale wastewater disposal and treatment field. It is hoped that these proceedings will be useful in partially mitigating this problem, and that they may attract the attention of those in government, in the private sector, and in academia to the need for continued work in this field.

February, 1982

Editors

Arild S. Eikum
Norwegian Institute for Water Research
Oslo, Norway

Robert W. Seabloom
University of Washington
Seattle, Washington, USA

SESSION I

OBJECTIVES OF ON-SITE WASTEWATER DISPOSAL

Chairman: T. Bilstad

OBJECTIVES OF ON-SITE WASTEWATER DISPOSAL

Robert W. Seabloom

Professor of Civil Engineering,
University of Washington, Seattle, Washington 98195,
USA

INTRODUCTION

Since ancient times man has disposed of his excreta and other wastes on the site where the wastes are generated. Probably the first disposal technology developed in response to a need for some kind of a management scheme to minimize the aesthetic problems, since early man most certainly did not perceive any connection between improper waste disposal and public health. Something akin to a crude pit-latrine probably was the first effort to minimize the offensive impact of indiscriminate discharge of human excreta on the land. Undoubtedly, the environmental and aesthetic insults must have become so obnoxious that the first aspiring sanitary engineers probably decided there must be a better way to manage the problem. Thus the pit-latrine or simple privy probably partially alleviated the nuisance and sufficed for many centuries. But as the population grew and man started to crowd together in urban areas, the problem of on-site waste disposal must have again become unmanageable. So, historically, sewer technology probably was developed in response to population density. Initially engineers concentrated on the collection and transportation phase using water carriage to remove the wastes

A. S. Eikum and R. W. Seabloom (eds.), Alternative Wastewater Treatment, 3–8.

from the densely population cities. "Out of sight out of mind" as it were. The concept of water carriage of human excreta then was a relatively new phenomenon. But the subsequent concentration of large volumes of urban wastewater containing human feces at a single discharge point gradually resulted in the gross pollution of the receiving waters. The theme "dilution is the solution to pollution" came into being, which in time led to many intolerable pollution problems in the receiving waters. A continued lack of environmental sensitivity in terms of the assimilative capacity of the receiving waters, along with cost limitations, aggravated these gross water pollution problems. Consequently, a whole new technology concerning wastewater treatment was developed relatively late in history some time in the latter 19th and early 20th centuries. Thus, in the developed countries we have seen in the last century the development of the central water carriage sewer system with the large centralized wastewater treatment facility and with discharge to the nearest receiving water. This rather sophisticated technology and philosophy thoroughly permeated the engineering educational constitutions and thus the thinking of a majority of the professional consulting engineers world wide. Only recently, and mainly in response to economic considerations has the engineering profession begun to take a hard look at this concept. As the comparative costs and environmental impacts became understood, it was necessary to consider small scale alternative wastewater disposal to reduce costs to an acceptable level.

SOIL AS A TREATMENT MEDIUM

In the developed countries the most common on-site wastewater disposal technique consisted of the septic tank soil-absorption system (ST-SAS). These ST-SAS relied upon the soil as a wastewater treatment medium. The soil is an exceedingly complex system, consisting of a solid, liquid, and gaseous phase, each of which possesses organic and inorganic constituents. Because it is an excellent habitat for microbial growth, and has the capacity for sorption of solute species, it has an enormous potential as a treatment medium. The soil particles provide an extraordinary amount of surface area for adsorption and ion exchange processes and allow for microbial digestion. Indeed the decomposition and recycling of nutrients is one of the primary functions of soil in the ecosystem. Hence when conditions are right, land disposal of wastewaters may be the most desirable option.

IMPLEMENTATION OF ON-SITE WASTEWATER DISPOSAL

The implementation or lack of implementation of on-site wastewater disposal systems has been influenced by numerous factors and decisions of many different parties.

1) On-site wastewater disposal systems were perceived as a temporary and inadequate disposal technique, mainly because a large number of the earlier ST-SAS were improperly sited, designed, constructed, and maintained and frequently failed giving the whole concept an unearned poor reputation.

2) Rules and regulations to manage and control the on-site wastewater disposal problem have evolved to be unnecessarily restrictive and biased against the concept, and often go far beyond what is considered reasonable for public health protection.

3) Special interest groups pushed for expensive gravity sewer systems with centralized treatment, while less costly on-site systems have largely been ignored.

4) The high technology solutions required large federal subsidies. With these federal subsidies the costs of water pollution control were not always borne by those who created the nuisance. For example about 25 percent of the U.S. population are not served by sewers, yet nowhere near that percentage of pollution abatement funds have been spent on rural sanitation.

5) Regulatory restrictions which imposed mandatory hookups to the centralized sewer system often mandated such a system.

6) Engineers and regulatory officials have generally reflected a traditional bias against small scale on-site wastewater disposal systems, plus a general lack of awareness of the cost and performance characteristics of such systems.

7) Partial blame for this lack of awareness results from the traditional engineering education which concentrates almost exclusively on the conventional systems.

8) When confronted with partial unfavorable site characteristics regulatory agencies frequently imposed sweeping building moratoria which were at times insensitive to local circumstances and often failed to take into account new technologies available.

9) The bias against on-site disposal techniques by engineers and designers has reflected a lack of understanding and belief that the concept is technically inferior as well as a managerial and administrative burden to the community. Thus the concept frequently receives only a token analysis in the water pollution abatement studies.

10) The extensive growth inducing characteristics so common to conventional gravity sewers are not generally a serious concern with on-site systems.

11) Engineers must recognize that properly sited, designed, constructed and maintained on-site systems are cost effective and in many instances are the preferred solution.

GLOBAL SANITATION PROBLEM

Of the four billion people on earth nearly three billion lack safe water supplies and proper excreta disposal, according to the eminent environmental engineering professor Abel Wolman (1). The lack of these facilities is probably responsible for 80 percent of the morbidity and mortality in the developing countries. This lack of sanitation becomes even grimmer when it is realized that children are particularly susceptible to the numerous water borne diseases that are prevalent in these areas. Half of the infants that die in the world each year, die from water borne diseases (2). The water supply and excreta disposal problems in developing countries are by no means confined to the rural areas. It is quite likely the most critical sanitation problems in the world are in the burgeoning shanty towns around the periphery of the rapidly growing third world cities. It has become obvious that the high technology solution, sewers and sewage treatment, often proposed by consulting engineers from the developed countries, is not cost effective. Thus there is a strong need to redirect these high technology engineers to think in terms of simpler solutions.

UNITED NATIONS, INTERNATIONAL DRINKING WATER AND SANITATION DECADE

In 1980 the United Nations launched the International Drinking Water and Sanitation Decade with the goal of bringing safe drinking water and proper human excreta disposal to all the world's people by the year 1980. While this goal, not yet reached in the U.S. and other developed nations, has been deliberately set high, Professor Wolman believes if you aim high, you will achieve more than if you aimed lower (1). If this goal has any hope of realization, engineers must propose a low technology approach, which consists of water supply and excreta disposal systems that the people can afford to build, operate and maintain. Estimates on the cost of achieving this goal range from US$ 140 to US$ 200 spread over the year period. However, G. Arthur Brown of Jamaica, the UN deputy administrator who is responsible for coordinating the inter-agency Water Decade efforts, puts it succintly "No matter how much it may cost to provide clean water and sanitation, the cost is far less than we are now paying for its lack."

CONCLUSIONS

Consequently, when considering the global problems, the objectives of on-site wastewater disposal becomes much broader, and can be broken down into the following two general categories:

For rural populations, developed countries;

- Provide safe wastewater disposal
- Provide cost-effective wastewater disposal
- Prevent aesthetic nuisances.

For rural and portions of the urban populations of developing countries

- Provide safe excreta disposal
- Provide cost-effective and safe wastewater disposal as needed.

Thus the purpose of this conference is to exchange research results as well as practical experience from small low cost wastewater treatment systems in the U.S. and several of the northern European countries.

REFERENCES

1. Dallaire, Gene: 1981, "U.N. Launches International Water Decade, U.S. Role Uncertain", Civil Engineering, March.
2. "Enfo", 1981, Environmental Sanitation Information Center Newsletter, Vol. 3, No. 2, June.

SOME ASPECTS ON NORWEGIAN PRACTICE

Lasse Vråle

Sanitary Engineer, Norwegian Institute for Water Research, Blindern, Oslo 3, Norway

INTRODUCTION

Approximately one million people out of Norway's 4 million inhabitants live in rural districts which makes it very difficult to connect to a municipal sewer. The solution of their wastewater disposal problem must be based on on-site wastewater disposal systems. In spite of the great number of Norwegians who need on-site disposal systems, there is still a long way to go before a good practical and permanent on-site disposal system is devised. To date very little work has been done on this important subject, and consequently there is a great need to improve the solutions presently in use in Norway.

Since this conference will deal with alternative low cost small wastewater treatment systems, it will be necessary to compare the on-site solutions with the conventional solution.

THE PURPOSE OF USING LOW COST SMALL WASTEWATER TREATMENT SYSTEMS

The purpose is to establish systems that treat the wastewater in such a manner that pollution of local surface water and groundwater is avoided. The treatment system must also have a

A. S. Eikum and R. W. Seabloom (eds.), Alternative Wastewater Treatment, 9–22.

low need for maintenance, and cost less than conventional systems. The cost should be low enough per system that it will not be necessary to build more houses in addition to the existing ones in order to finance the wastewater treatment system. Thus it is necessary that solutions are found that make it possible to maintain the same type and density of dwellings without creating new pollution problems.

NORWEGIAN LAWS AND REGULATIONS CONCERNING WATER SUPPLY AND WASTEWATER TREATMENT MAKE IT EASY TO USE SEWAGE PROBLEMS AS THE MOST EFFICIENT TOOL TO REGULATE BUILDING DEVELOPMENT

The Existing Norwegian regulations make it easy for the authorities to use wastewater management to obtain a desired home building policy. For example, it is possible to refuse building permits in certain districts for various reasons and to stimulate house-building in other areas. Many Norwegian communities have since the last war had accelerated building programs in order to meet the housing demand. The most efficient way to increase housing production is to build a number of houses at the same time, connected to a collection system and one treatment plant. The cost of sewers and water mains has increased because the construction is much more difficult in rocky terrain. As a result, to keep the cost relatively reasonable per dwelling, it has become necessary to use smaller lots and denser developments.

Thus when Norwegian authorities seem restrictive against on-site wastewater disposal systems, it does not necessarily mean that they do not believe in the systems, or that the systems don't work, but rather that the authorities want a conventional solution to get a higher building density in the district.

Another problem is that permanent on-site disposal systems may create problems for the establishment of a conventional system later on. Therefore it is common to have people who are allowed to build an on-site disposal system, sign a document stating that they will connect to the municipal sewer if and when it is constructed in their district. It is the writer's feeling that this has slowed down the development of good permanent on-site disposal systems in Norway.

NORWEGIAN INVESTIGATIONS SHOW THAT MANY ON-SITE TREATMENT PLANTS WORK RATHER POORLY

Since spring 1975, several investigations of on-site wastewater treatment systems have been performed. The results of the first three investigations were published in 1977 and were undertaken in Aust-Agder county (1), in Vestfold county (2), and in parts of Akershus county (3). These investigations resulted in the following conclusions:

- A high number (80-90 percent) had serious construction faults
- The maintenance of the treatment plant was neglected
- The removal of sludge was neglected
- The state regulations must be simplified and easier to understand
- The information about the treatment plants and the skill of the people that construct the plants must be better.

The problems observed were due more to incorrect construction, poor control and operation rather than on-site processes itself. However, there is no doubt that the Norwegian practice of on-site wastewater disposal can be improved. The type of problems reported seems to be similar to what has been previously found at conventional wastewater treatment plants.

A CHANGE IS NECESSARY IN THE NORWEGIAN POLICY CONCERNING ON-SITE WASTEWATER DISPOSAL SYSTEMS

In the author's opinion, it has been too easy in the past to use water and wastewater regulations as a tool to control building programs. In addition, conventional systems seem to enjoy a much better reputation in relation to on-site disposal systems. A building permit was easily obtained if it was possible to connect the sewer to an existing municipal sewer. No one asked or seemed to care whether there was adequate wastewater treatment at the other end of the sewer. As long as wastewater from the new house disappeared, nobody worried whether the wastewater ended up at the treatment plant or not, and consequently it was very difficult to advocate on-site disposal systems.

ON-SITE GROUNDWATER USE FOR WATER SUPPLY MUST GET A HIGHER PRIORITY

Groundwater use for local water supply is also an important consideration when on-site wastewater disposal systems are proposed.

Frequently unrealistic, high estimates of future water demand in Norway have been made, as shown in Figure 1, which shows the difference between the estimate of water demand in Nesodden municipality, worked out in 1967, and the real water use delivered from the water treatment plant. The reduction illustrated by the curve was due principally to a leakage control program, and a 50 percent reduction in water use. Out of Nesodden's 9,800 inhabitants, 5,000 are served by municipal water while the rest have local water supply from wells. Due to the high estimates of water demand, it was believed that local wells and water supply were not sufficient. It was decided that all inhabitants had to base their water supply on municipal delivery, and houses had to be connected when the watermains were constructed.

THE SIZE OF THE ON-SITE DISPOSAL SYSTEMS IS BASED ON WASTEWATER FLOW

It is realistic to design on-site systems for houses with and without water closet. Experience has shown that the water

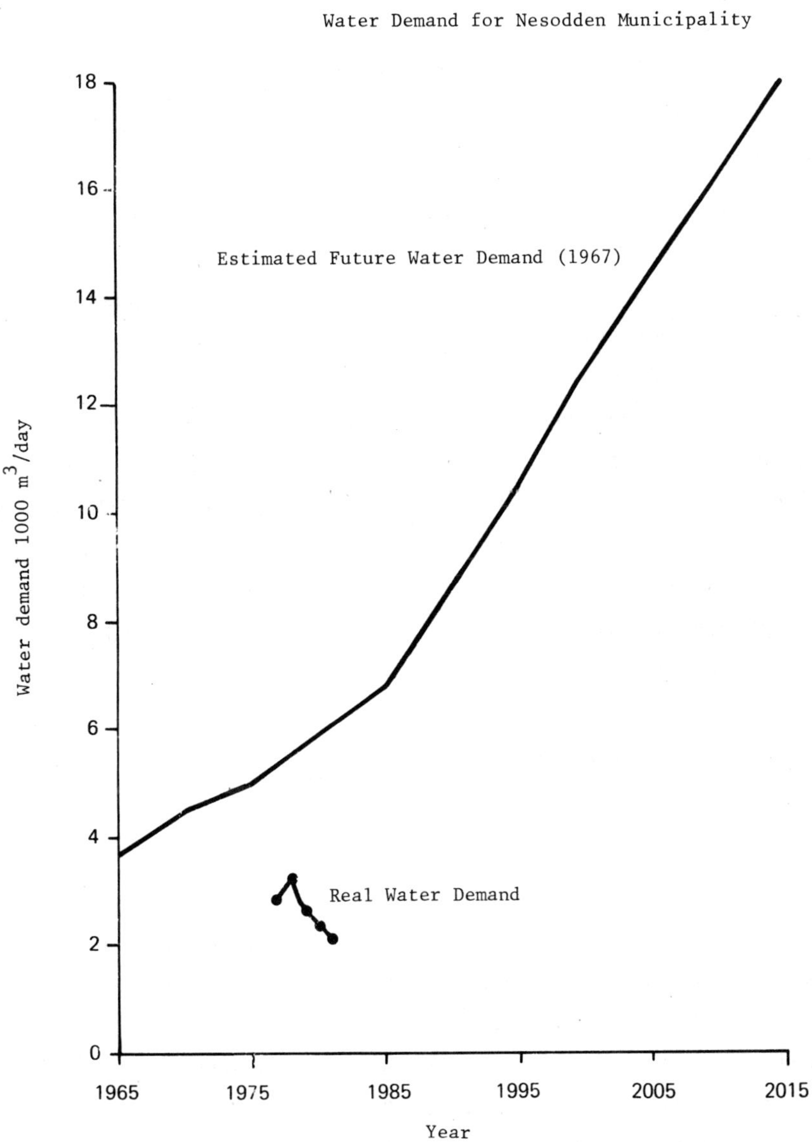

Figure 1. Estimated Future Water Demand worked out in 1967 for Nesodden Municipality (4).

demand for single houses can be very different with different installations. American information concerning water demand for single houses shows a higher demand than for Norwegian houses in rural districts (5). However, more information is needed on this important subject.

There is a need to be more concerned about water consumption and the amount of wastewater generated, because the wastewater flow directly determines the size of the on-site disposal system. For example, Norwegian single family houses without water closets have been found to use as low as 100 liter/day. Traditionally, 200 l/p.d has been used as a recommended value for water demand in Norwegian households. This is probably too high. An investigation of water consumption in single family houses without water closet at Danskerud (6) in Ås municipality showed an average water consumption of 83 l/p.d. Some of the solutions for on-site treatment systems or small treatment plants may also be used on a somewhat larger scale by connecting several homes. It is important to note that many of the conventional treatment processes that have been scaled down, do not work very well with very low flows.

WHAT IS LOW COST ON-SITE WASTEWATER DISPOSAL?

Cost is relative. In many cases the use of alternative on-site wastewater treatment systems is the only possible way to get a building permit. The house may be located many kilometers from the municipal sewer system. So if the authorities allow a suggested plan, the owner will have to decide whether to assume the cost or drop the building plans. Since the authorities often do not trust on-site systems, they will insist on a conventional system with collection and treatment in a larger plant, no matter what it costs. It is a common situation in Norwegian communities, when the proposed house is located some distance away from the nearest sewer, the authorities will only allow the building permit after the owner agrees to connect to the municipal sewer.

The author knows of one situation where people had to pay NOK 300,000 (US$ 50,000) in order to connect a single house to the municipal sewer.

Another example would be where several existing houses were located considerably apart from one another on rather large lots (5,000 to 15,000 m^2). Frequently, many water closets may have been installed in these homes without permits since on-site treatment systems were not allowed in the community. Not too surprisingly, the well water supply became contaminated and the residents then invited the municipality to suggest a solution to alleviate the situation. The design suggested by the authorities usually consisted of a conventional sewer system and wastewater treatment plant. The typical placement of the storm sewer, sewage, and water pipes is shown in Figure 2. Even though a municipal water supply may be years away, for safety reasons, the water main is placed above the sewer and storm sewer, and since the water main must be laid below the freezing level, it means blasting will be necessary in rocky terrain. The cost of this solution is of course rather high, frequently amounting to NOK 300,000 - NOK 500,000 per house and certainly is not an acceptable solution. In order to bring the cost down to the normal NOK 50,000 - 60,000 per house, 4 to 6 times the number of houses must be built to justify the recommended solution. So, instead of taking care of sewage for 40 houses, it is necessary to treat sewage from 230 houses. This is commendable if the goal was to build as many houses as possible, but if the goal was to minimize a pollution problem and maintain the environment, the solution was poor. Thus there is a great need for cost effective on-site disposal systems that will be accepted on the same basis as conventional systems.

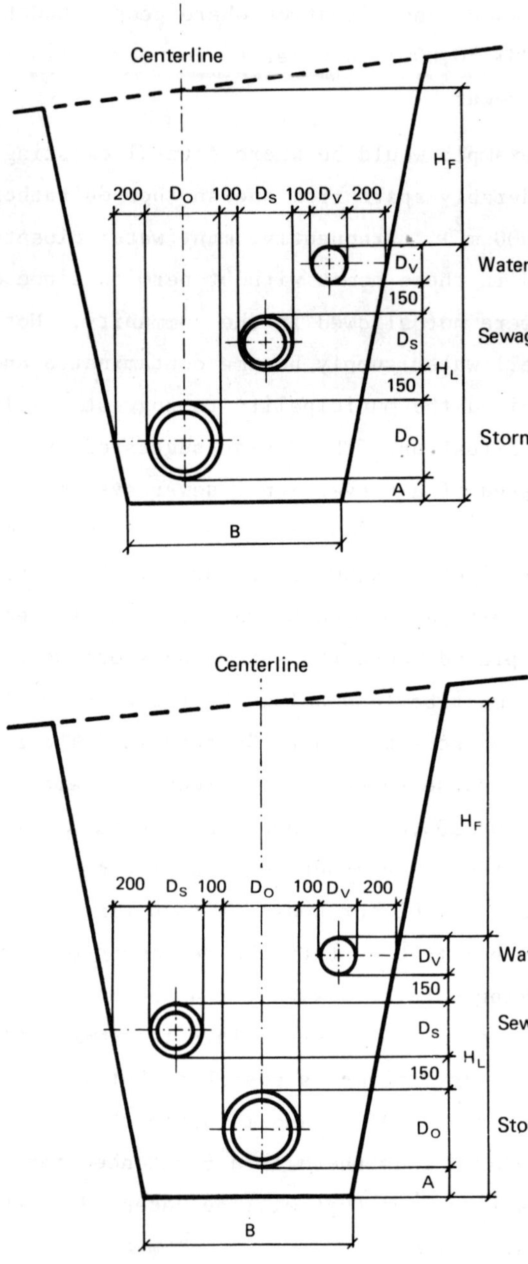

Figure 2. Typical arrangement of two-pipe separate sewer system in Norway.

CHEAPER COLLECTION SYSTEMS WILL MAKE IT POSSIBLE TO COLLECT SEWAGE FROM EXISTING DWELLINGS WITHOUT HAVING TO INCREASE THE HOUSE DENSITY IN ORDER TO FINANCE THE COLLECTION SYSTEM

The following solutions are suggested to bring down the cost of collection systems in such districts:

1. Eliminate the storm sewer, use local infiltration of rain water.
2. Eliminate the water main, use local water supply.
3. Use shallow trenches and insulate the pipe where necessary.
4. Use low pressure sewer systems (LPS), as shown in Figure 3.
5. Use tank collection system where other systems are not possible.

WHAT POLLUTANTS SHOULD BE REMOVED BY ON-SITE WASTEWATER TREATMENT SYSTEMS?

This is a very important question the answer to which must be clear before the most effective treatment system can be chosen. For conventional treatment plants in Norway, phosphorus has been the most important pollutant to be removed followed by organic material as measured by BOD or COD. For small treatment plants and on-site disposal systems, the same water quality parameters have been adopted to measure performance.

Self-purification in receiving waters and phosphorus-binding in soil must also be taken into consideration. In rural districts the non-point pollution load will be relatively high, compared to the wastewater load from dwellings. In other countries, more emphasis has been put on removal of BOD, ammonia, total nitrogen, and E-coli. Removal of phosphorus is difficult in most commonly used units in on-site treatment systems, although the necessity for its removal needs to be documented.

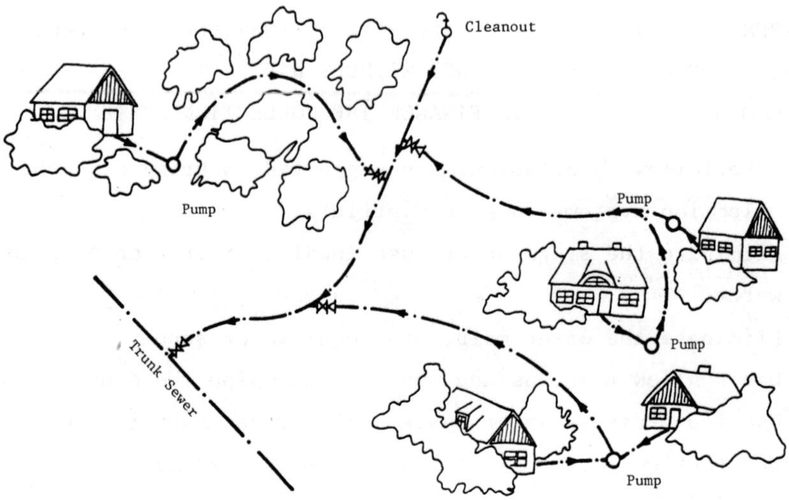

Figure 3. Low Pressure Sewer System, LPS.

WHAT DEGREE OF POLLUTANT REMOVAL IS NECESSARY FOR ON-SITE SYSTEMS IN ORDER TO COMPETE WITH CONVENTIONAL SYSTEMS?

First, it will be necessary to find out what degree of pollutant removal conventional plants have provided. Since 1976 NIVA has examined most conventional treatment plants in Norway and established that up to 50 percent of the treatment plants were not functioning as they should. The main reasons for the poor performance were the following:

1. Poor collection systems with high inflow during rainy weather. This creates shock loading which disturbs the process.
2. Failure in construction and installations.
3. Poor operation of the treatment plant.
4. Lack of knowledge about the treatment processes and necessary treatment routines among treatment plant operators.

These results were based upon one single visit to each plant. Experience has shown, especially for smaller conventional treatment plants, that the treatment systems may work poorly for long periods of time, because the basic process may make it easy for the wastewater to pass the treatment plant independent of the degree of removal. This is not possible for the filtration process since poor maintenance and treatment will clog the filter and the water will back up. In addition, high hydraulic loadings may cause serious failure of performance for the activated sludge process, due to sludge loss. Treatment plants with trickling filter or rotating discs may take the hydraulic shocks better than the activated sludge processes. Chemical treatment plants have the highest performance, although they are dependent on a stable and correct addition of chemicals.

Another very important factor is the degree of collection of sewage. The larger the treatment plant is, the larger and more complicated the collection system will be. Investigations have shown that the degree of collection of sewage (5) is rather low for many treatment plants in Norway, and an average amount of 50 percent has been found. Table 1 shows the result from this investigation. The reason for this surprisingly low degree of collection is:

1. The collection system has not been completed at the same time as the treatment plant.
2. The pollution control authorities do not have a clear requirement for collection of the sewage.
3. Older sewers/storm sewers with open joints built for other reasons than collection of sewage, are used as a part of the system to save money. These old sewers leak both ways; sewage leaks out of the system, and rain water seeps in.

The degree of collection is an important factor when solutions for on-site wastewater treatment are discussed. It is easier to collect all of the wastewater coming from a single residence by way of a short sewer, than when the wastewater has to travel relatively long distances and possibly pass several pumping stations.

Table 1. Degree of Collection for some Norwegian Treatment Plants.

Report No.	Municipal sewer district	Time for measurement	Pollution load to treatment plant (average) kg P/d	kg N/d	Degree of collection: Municipal sewer district %	Collection system %
1	Hoffsveien	Fall 1977	17.3	8.3	67	-
2	Nordre Follo	1976	(19.4)	(103)		53
3	Skarpsno	1977	66.6		55	-
4	Slemmestad,Asker	1977	1.8	16.4	10	-
5	Slemmestad,Røyken	1977	4.7	25.4	43	-
6	Notodden	Summer 1979	(10.1)		-	(31)
7	Harestua	1976-78	1.2		-	(55)
8	6 treatment plants Nittedal	1977-78	various		-	(80)
9	Raufoss	Summer 1979	8.5		50	70
10	Lillehammer	1979	38.4		?	75
11	Hamar	Summer 1979	-			
12	Gjøvik	Summer 1979	-			
13	Hønefoss, Monserud	Jan.1979-Jan.1980	8.7	39.9	15	57
14	Nesbyen	1979			-	85
15	Mjøndalen	1979				80
16	Bekkelaget	1978		(47)		
			Arithmetic average		40	79
			Weighed average		44	68

When the degree of removal is to be compared between on-site systems and conventional systems, the conventional systems must be corrected for a lower degree of collection, as shown below:

	Average degree of removal %	Degree of collection %	Over all effect Degree of removal %
Conventional system	75	60	45
On-site disposal system	60	100	60

WHAT IMPROVEMENTS ARE NEEDED IN NORWEGIAN PRACTICE FOR ON-SITE WASTEWATER DISPOSAL SYSTEMS?

Small wastewater treatment systems and on-site disposal systems must be accepted on the same level as larger and more conventional systems. However, more information is needed on how the different systems work in practice, how much they cost, along with the degree of removal and the degree of collection they provide in comparison to conventional systems. The technical solution should be chosen only after cost-benefit analyses utilizing local information have been carried out.

The control and the degree of removal for conventional treatment plants in Norway must be improved. It is also important that the central agency require all homes in the municipal sewer district to be connected to the sewer, in order that all of the sewage ends up at the treatment plant. This will be costly, and the municipality will likely be forced to use more economical collection systems. Municipalities will also be forced to operate with more realistic sewer districts, which means smaller districts. When the sewer districts are reduced in size, the need for low cost small wastewater treatment systems in the peripheral districts will be greatly increased.

It is important to find technical solutions that make it possible to elevate the infiltration pipe lines. Presently, frost regulations require a depth of 1.60 m below surface. This limits the use of natural infiltration due to bedrock and high groundwater table, since the upper part of the ground is best suited for infiltration. The mound principle should be allowed in the new Norwegian regulations which are now being formulated.

REFERENCES

1. Andreassen, E.: 1977, "Kritisk vurdering av infiltrasjonsanlegg i spredt bebyggelse". (Critical Evaluation of Infiltration Systems in rural Areas.) VANN No. 4.

2. Skudal, L.: 1977, "Undersøkelse av avløpsanlegg for spredt bebyggelse i Vestfold". (Investigation of Wastewater Treatment System in Rural Areas in Vestfold.) Vestfold fylkeskommune - Planavdelingen, Dated Sept. 21.

3. Smits, C.: 1977, "Vedrørende undersøkelse av 84 private renseanlegg for eneboliger i Fet, Lørenskog, Nittedal og Rælingen. Tilstandsbeskrivelse. Årsaker til utilfredsstillende forhold. Forslag til Tiltak".(Investigation of 84 private treatment plants for single houses at Fet, Lørenskog, Nittedal and Rælingen. Description of the present situation.) Avløpssambandet Nordre Øyeren (ANØ).

4. Bjelke, W.: 1967, Rapport Nesodden kommune. Vannforsyning Sak 1429. (Report Nesodden Municipality. Water Supply Case 1429.) Erik Ræstad A/S.

5. Otis, R.J.: 1980, "Mound Systems". Design Manual Onsite Wastewater and Disposal systems. U.S. Environmental Protection Agency EPA 625/1-80-012.

6. Skaarer, N.: 1981, "Vannforbruk i boliger uten vannklosett langt lavere enn ventet". (Water Consumption in Dwellings without Water Closet far less than assumed.) BYGG 5 A, Teknisk Ukeblad, June 19.

7. Vråle, L.: 1981, "Virkningsgrad for oppsamlingsnett. - Status for eksisterende målinger". (Efficiency of Collection Systems. Status of Existing Measurements.)

SESSION II

QUANTITY AND CHARACTERISTICS OF RESIDENTIAL WASTEWATER

Chairman: T. Bilstad

ALTERATION OF IN HOUSE WASTEWATER FLOW WITH LOW FLUSH TOILET FIXTURES AND GRAYWATER RECYCLE

W.C. Boyle, Professor
R.L. Siegrist, Research Specialist
D.L. Anderson, Project Specialist

Department of Civil & Environmental Engineering,
University of Wisconsin, Madison, Wisconsin 53706, USA

INTRODUCTION

Clean water has been regarded as an unlimited resource for many years, and little consideration has been given to the cautious, conservative use of water. However, the problems and impacts associated with water consumption are stimulating interest in water conservation. Alleviating overloads to water supply and sewage treatment facilities, reducing operational requirements and costs, as well as rendering innovative technologies more feasible are being recognized as potential benefits of reducing water demands and wastewater loads.

A significant portion of consumptive water use and concommitant wastewater flow in the U.S. results from domestic activities. As of 1975, domestic uses accounted for more than 40 percent of the water use that produced wastewater requiring subsequent treatment (1). Numerous activities occurring inside and outside a residence contribute to this water demand. Of particular importance are those activities which not only consume water, but also produce wastewaters that require treatment and disposal; namely, the interior water-using activities. Toilet

A. S. Eikum and R. W. Seabloom (eds.), Alternative Wastewater Treatment, 25–42.

usage, bathing, and clotheswashing account for more than 75 percent of the interior water use and wastewater flow from U.S. residential dwellings (Table 1).

Table 1. Average Residential Waste Flow Characteristics[a]

Activity	L/use	Use/Cap/d	Daily Flow L/Cap/day[b]	Daily Flow %
Toilet Flush	16.3	3.5	61.2	35
Bathing	92.6	0.4	34.8	20
Clotheswashing	141.4	0.3	37.8	21
Dishwashing	33.3	0.3	12.1	7
Garbage Grinder	7.6	0.6	4.5	3
Miscellaneous	-	-	24.9	14
Total	-	-	172.4	100

(a) Based on results reported in ref. (2) - (6).

(b) L/Cap/day may not equal L/use x use/Cap/day due to differences in number of study averages used to compute means shown.

Strategies to achieve waste flow reduction are numerous and varied. Consumer education to yield improved water-use habits, the use of water-saving devices, fixtures and appliances, and wastewater recycle systems are several alternatives that can be used to produce waste flow reductions. Of these, the use of water-saving devices, fixtures and appliances, appears to offer the greatest potential at present. Furthermore, since the major component of residential wastewater comes from the toilet (35 percent of total wastewater produced), significant water flow reductions can be achieved through the use of innovative plumbing systems.

To enhance the limited data base available regarding the performance of these innovative toilet systems, a field demonstration project was performed at the University of Wisconsin under sponsorship of the U.S. Office of Water Research and Technology. This study evaluated the performance of extreme low flush toilet fixtures and wastewater recycle systems with the objectives being to investigate the impacts of these units on water use and waste flow, to delineate the installation, operation, and maintenance requirements, to determine user acceptance, and to estimate costs. The methods and results of this study are documented elsewhere (7), and only a brief synopsis is presented herein.

MATERIALS AND METHODS

At the onset of this project, a search was made to identify commercially available low-flush toilet systems and wastewater recycle systems. The three low-flush toilet systems which were identified and selected for this study are listed in Table 2. Data in this table were based upon manufacturer's literature and communications with manufacturer. All were similar in appearance to a conventional U.S. water closet, but required flush volumes of only 1.5 to 6.0 L. To enable flushing with such small volumes, the Microphor employed a burst of compressed air, the Monogram utilized a small macerator pump, and the Ifo possessed a blowout action. A total of nine fixtures were installed in five typical American homes, and monitoring of their performance occurred over a total of 1926 days. Usage data were collected automatically for interior water use, fixture use, and toilet water use with specially altered water meters and flow indicating switches interfaced with occurrence counters and continuous strip-chart recorders. Data were collected at each home over many multi-day periods and analyzed on an hourly and daily basis. System operations were monitored with separate power meters and elapsed

Table 2. Characteristics of Extreme Low-Flush Toilets.

Characteristic	Toilet Microphor	Monogram	ifo
Flushing mode	Water assisted by compressed air	Water assisted by macerator pump	Water
Flushing actuation	Depress/release lever	Depress/release bottom	Lift/release handle
Resource requirements			
Water	1.9 L/use	1.5 L/use	6 L/use 3 L/use
Power	yes	yes	no
Application considerations	Compressed air required; Standard rough-in; water pressure 20-50 psi	.Power supply standard rough-in Water pressure ≥ 20 psi	Non-standard Rough-in Drain line slope ≈3%
Maintenance	Service Compressed air source	Grinder replacement at 2-5 yr.	---
Development state	Fully developed	Prototype	Fully developed
Manufacturer	Microphor, Inc. Willits, Cal.	Monogram Industries, inc. Long Bearl,Cal.	Ifo Sanitaire AB S-29500 Bromolla, Sweden
Approximate cost**	$410 plus compressor	$1,000∔	$240-365

**Shipping cost additional to cost presented.
∔Prototype cost shown. Manufacturer's predicted production model cost is approximately $500.

time indicators, and maintenance requirements were delineated in a log. User acceptance was assessed through recordings in a user log and responses to a final questionnaire.

Three commercially manufactured home recycle systems that purified bathing and laundry washwaters for reuse in toilet flushing were initially selected for study, but only the Aquasaver system proved to be truly available. This system employed the unit processes of sedimentation, pressure cartridge filtration (20 μm), and chlorine disinfection in a small inhouse module. Two recycle units were installed in two homes and monitoring of their performance occurred over a total of 559 days. Monitoring during this period was similar to that for the low-flush systems, but also included other flow streams associated with the recycle systems. Qualitative characterization was also accomplished through collection of grab and 24-hour flow composited samples with analyses for various parameters performed according to Standard Methods (8).

Regulatory and public information agencies were contacted to locate at least four residential dwellings interested in participating in this study. Efforts were made to include typical dwellings with normal water, using fixtures and one or more children in the residing family. After reviewing interested participants, five sites were selected as primary study residences. Characteristics of these study residences appear in Table 3. Note that residence VA and VAR are the same household, used for both low-flush toilet and recycle system study.

Table 3. Characteristics of Study Residences.

Characteristic	Residence					
	VA	MI	GA	GL	VAR	NE
Fixture Unit	Microphor	Monogram	Ifo	Ifo	Aquasaver	Aquasaver
Installation	New	Retrofit	Retrofit	New	New	Retrofit
Residents						
Adults	2	2	2	2	2	2
Children	2(3,5)	2(5,8)	1(1)	4(1,2,9,10)	3(4,6,14)	2(17,19)
Dwelling floors	2	1	2	2	2	1
Bedrooms	3	3	4	3	3	3
Toilets	2	1	2	2	2	1
Shower	yes	yes	yes	yes	yes*	yes
Bathtub	yes	yes	yes	yes	yes*	yes*
Bath sinks	yes	yes	yes	yes	yes*	yes
Auto.clothes wash	yes	yes	yes	yes	yes*	yes*
Auto dishwasher	no	yes	no	yes	no	no
Lot size(hectares)	7.7	.06	4.9	1.2	7.7	0.14
Water supply	well	city	well	well	well	well
Waste Disposal	Holding tank	City	S.T. drain-field	Holding tank	Holding tank	S.T.drain-field

RESULTS

Water Use and Wastewater Flow

The inhouse water use and waste flow characteristics measured in this study are tabulated in Table 4. At the four homes with the extreme low-flush toilets, the average daily flow ranged from 42.4 to 123.0 Lpcd, of which 6.0 to 19.6 percent was due to toilet flushing. At the three homes that exhibited the low per capita consumption, conventional water-saving faucets and showerheads (9-12 Lpm approximately) were also present. Day-to-day variations about the mean day at each home were similar, with the lower and upper limits of the 95 percent confidence interval for total daily flow varying from 78 to 88 percent and 112 to 123 percent of the respective mean daily flows. The maximum daily flow experienced at the homes ranged from 170 to 350 Lpcd, or from 285 to 402 percent of the respective mean daily flows. Hourly flow data indicated wide fluctuations with minimum flows of zero and maximum flows ranging from 44.2 to 94.6 Lpch (Table 4). Increases in water use above the mean day and hour were found to be closely correlated with increases in the non-toilet use component of the flow.

At the two homes with the washwater recycle systems, the average daily freshwater flow varied from 78 Lpcd to 127 Lpcd (Table 4). Day-to-day variation, shown by the limits of the 95 percent confidence interval for daily flow expressed as a percent of the mean day, were 75 and 125 percent to 89 and 111 percent. Since toilet water use was recycled washwater, freshwater use for flushing was essentially eliminated as long as washwater production and storage met toilet flushing demands. At home NE, no additional freshwater was needed for flushing, while at home VAR, an average of 2.3 Lpcd was required. This was due to occasional unfavorable balances between available washwater and required flushwater, as well as to a slight difference in recycle system

Table 4. WATER USE AND WASTEFLOW CHARACTERISTICS*

Parameter		Extreme Low-Flush Toilets				Washwater Recycle	
		Microphor [VA]	Monogram [MI]	Ifo [GA]	Ifo [GL]	Aquasaver [NE]	Aquasaver [VAR]
Interior Water Use	Mean	75.2	123.0	78.6	42.4	126.7	78.0
	S.D.	52.8	84.0	46.3	33.5	58.1	68.1
	95%-	66.2	103.9	63.9	32.8	112.7	58.8
	C.I.	84.3	142.1	93.0	52.0	140.6	97.2
	Min.	14.2	28.4	25.2	6.3	47.3	15.1
	Max.	298.1	350.1	239.7	170.3	364.3	280.1
	Max hr.	75.6	94.6	63.1	44.2	113.6	68.1
Toilet Water Use	Mean	4.5	11.6	9.8	8.3	33.1**	24.7**
	S.D.	2.2	2.5	6.0	3.0	18.2	10.7
	Min.	1.0	4.4	1.0	2.6	9.4	7.3
	Max	15.6	18.7	28.3	16.8	106.2	49.8
	Max.hr.	4.2	3.3	7.1	4.4	22.0	22.3
Non-Toilet Water Use	Mean	70.7	111.4	68.9	34.2	126.7	75.7
	S.D.	52.5	84.0	45.1	32.8	58.1	67.6
	Min.	9.6	18.5	18.9	1.9	47.3	14.2
	Max.	293.8	336.9	230.6	157.1	364.3	274.2
	Max.hr.	75.2	93.0	63.1	43.3	113.6	68.1
Fixture	Mean	2.54	6.02	1.95	1.56	2.65	2.00
	S.D.	1.23	1.35	1.22	0.57	1.46	0.86
	Volume	1.9	1.9	3.0/5.3	5.3	12.5	12.3
Data Pts.	Total	133	77	41	49	69	51
Residents	No.+	4(3,5)	4(5,8)	3(1)	6(1,2,9,10)	4(17,19)	5(4,6,14)

*Results presented in Lpcd except max hr flow (Lpch), fixture usage (npcd), max hr fixture usage (npch), and flush volume (L)
**Toilet water use is recycled washwater; not freshwater
+Total residents with children's ages (yr.) in parentheses.

design requiring freshwater for chlorine feed with each use. The maximum daily freshwater flow at the two homes was 280 Lpcd and 364 Lpcd (Table 4). Hourly water use data again indicated wide fluctuations with minimum flows of zero and maximum flows of 68.1 and 113.6 Lpch. Maximum day and hour flows were due almost entirely to non-toilet water use at these sites.

Water Quality

The quality characteristics of various process streams associated with the recycle system are shown in Tables 5 and 6. Appreciable amounts of pollutants were found in the raw graywater, confirming the results reported by previous investigators (3,9, 10). The recycle water quality varied widely and was generally of a much lower quality than a typical freshwater toilet supply. Variations in coliform bacteria concentrations in the recycle water corresponded closely to variations in residual chlorine levels. As long as a measurable chlorine residual was present, total and fecal coliform values were less than 10 org./100 mL.

Only limited pollutant removals were achieved by the recycle system. At home NE, pollutant reductions ranged from 0 to 70 percent. The increase in total solids from raw to recycle was expected due to the addition of calcium hypochlorite in the disinfection process. Results at home VAR revealed increases in many pollutant levels. The reason for this was not clear, but day-to-day variability in the raw washwater and attenuation in the storage tank may have been responsible. At both homes, 20 μm pressure cartridge filtration afforded minimal pollutant removals, with the bulk of any renovation occurring as a result of the sedimentation and disinfection processes.

The quality of the recycled water is important for health, safety, and operational considerations as well as from an aesthetic viewpoint. While no regulatory standards exist regarding recycle water quality for home toilet flushing and lawn irrigation, the National Sanitation Foundation (NSF) has set standards associated with testing and certification of home recycle systems (11,12). According to NSF Criteria C-9, recycle water must possess Total Coliforms $\leq$ 1/100 mL, turbidity $\leq$ 100 TU, BOD_5 influent, and TSS $\leq$ 90 mg/L. An Aquasaver was tested under lab conditions by NSF and approved per these criteria.

Table 5. Physical-Chemical Water Quality, Residence NE.

PARAMETER	UNITS	STATISTIC	RAW WASHWATER	STORAGE EFFLUENT	RECYCLE WATER
BOD_5	mg/L	Mean (n)	125 (7)	100 (6)	74 (10)
		S.D.	52	28	25
		Range	33-193	52-130	34-110
COD	mg/L	Mean (n)	242 (5)	204 (6)	216 (7)
		S.D.	73	46	28
		Range	169-355	123-259	171-256
TS	mg/L	Mean (n)	794 (6)	950 (7)	1040 (9)
		S.D.	168	114	172
		Range	626-1076	770-958	860-1400
TVS	mg/L	Mean (n)	128 (6)	166 (7)	172 (8)
		S.D.	48	59	69
		Range	62-180	126-262	124-324
TSS	mg/L	Mean (n)	36 (4)	18 (4)	12 (5)
		S.D.	14	7	3
		Range	21-55	13-28	8-15
TVSS	mg/L	Mean (n)	33 (4)	15 (4)	10 (5)
		S.D.	11	5	3
		Range	20-47	11-22	6-15
TKN	mg/L	Mean (n)	5.8 (6)	4.8 (7)	5.2 (10)
		S.D.	2.7	2.8	3.8
		Range	2.0-10.1	1.5-9.2	0.3-10.8
NH_4-N	mg/L	Mean (n)	0.6 (4)	0.5 (4)	0.2 (6)
		S.D.	0.6	0.8	0.3
		Range	0.35-1.4	0.1-1.6	0-0.7
NO_2-NO_3-N	mg/L	Mean (n)	0.5 (4)	0.5 (3)	0.4 (5)
		S.D.	0.5	0.2	0.4
		Range	0.3-1.23	0.35-0.67	0-0.9
Total P	mg/L	Mean (n)	1.0 (6)	2.8 (7)	1.0 (11)
		S.D.	0.5	3.6	0.6
		Range	0.08-1.7	0.18-10.5	0.2-2.1
TURB	NTU	Mean (n)	42 (5)	42 (6)	36 (9)
		S.D.	13	6	6
		Range	26-55	30-47	27-41
pH	-	(n)	-		13
		Range	-		7.0-7.8
Total Avail. Cl_2	mg/L	Mean(n)	-		4.2 (15)
		S.D.	-		6.1
		Range	-		0-19
Fecal Coli	Log no./ 100/mL	Mean (n)	- (5)	-	-(8)
		Range	1.90-7.34	-	0-<2.7

Table 6. Physical-Chemical Water Quality, Residence VAR.

PARAMETER	UNITS	STATISTIC	RAW WASHWATER	STORAGE EFFLUENT	RECYCLE WATER
BOD_5	mg/L	Mean (n)	147 (6)	200 (6)	185 (8)
		S.D.	45	89	80
		Range	80-215	60-319	58-317
COD	mg/L	Mean (n)	276 (5)	389 (6)	383 (8)
		S.D.	56	134	133
		Range	230-359	136-511	201-577
TS	mg/L	Mean (n)	810 (6)	1143 (7)	1108 (9)
		S.D.	215	248	298
		Range	623-1086	834-1503	536-1496
TVS	mg/	Mean (n)	179 (6)	282 (7)	259 (9)
		S.D.	56	93	91
		Range	117-259	142-394	92-369
TSS	mg/L	Mean (n)	92 (6)	71 (7)	66 (9)
		S.D.	64	24	31
		Range	39-211	40-100	27-124
TVSS	mg/L	Mean (n)	38 (6)	38 (7)	36 (9)
		S.D.	19	24	15
		Range	16-68	12-86	17-56
TKN	mg/L	Mean (n)	5.7 (5)	9.8 (3)	9.1 (5)
		S.D.	2.4	6.3	4.8
		Range	2-8	5.4-17.0	4.0-16.0
NH_4-N	mg/L	Mean (n)	1.2 (4)	1.8 (2)	1.6 (3)
		S.D.	1.1		1.7
		Range	0-2.6	1.4-2.2	0-2.0
NO_2-NO_3-N	mg/L	Mean (n)	0.4 (3)	1.1 (2)	1.2 (4)
		S.D.	0.2		0.8
		Range	0.2-0.6	0.2-2.0	0.2-2.0
Total P	mg/L	Mean (n)	0.3 (2)	-	-
		S.D.	-	-	-
		Range	-	-	-
TURB	NTU	Mean (n)	-	-	-
		S.D.	-	-	-
		Range	-	-	-
pH	-	(n)	-	-	6
		Range	-	-	6.8-8.2
Total Avail. Cl_2	mg/L	Mean (n)	-	-	4.7 (11)
		S.D.	-	-	7.8
		Range	-	-	0-25
Fecal Coli	Log no./ 100 ml	Mean (n)	- (5)	-	- (6)
		Range	5.04-7.78	-	2-4.78

Subsequently, Standard 41 was developed requiring Total Coliforms $\leq$ 240/100 mL, turbidity $\leq$ 90 TU, $BOD_5 \leq$ 45 mg/L, TSS $\leq$ 45 mg/L, and a nonoffensive odor (12). Under field conditions in this study, the Aquasaver recycle water met NSF Standard 41 limits, except BOD_5 at both homes and TSS at home VAR.

Impacts of System Use on Water Use Characteristics

Assessing the impacts of water saving technologies on the water use characteristics at a given home can be a complex task. At first glance, the most direct method would appear to be a comparison of data obtained prior to installation of the water-saving technologies to post-installation data. However, this approach suffers from potentially serious shortcomings. For new homes, this comparison is impossible. At existing homes, changes in water demand and fixture usage totally unrelated to the water-saving technologies employed, can yield inaccurate results. These changes in water use can be due to rapidly changing life-styles of growing children, changing work schedules and lifestyles of adult residents, and changes in the physical characteristics of the dwelling including water-using fixtures, water supply pressure, and so forth. At the three homes where background data were collected, changes in water demand and fixture use with time were exhibited that were apparently unrelated to the innovative toilets. An alternative strategy utilizes actual measured fixture usage data. For the innovative toilet systems, the measured toilet usage data with these systems were utilized to determine the water use characteristics that would have occurred if conventional toilets had been used instead. These water use data were then compared to the measured data to assess the impacts of the innovative toilets. The results of this analysis follow.

The use of low-flush toilets rather than conventional fixtures provided major reductions in daily water use (Table 7). Compared to a 13.2 L toilet, reductions of 12 to 68 Lpcd, or 17 to 36 percent were calculated, while compared to an 18.9 L toilet, reductions of 21 to 102 Lpcd, or 26 to 45 percent were calculated. The use of the low-flush toilets did not affect routine day-to-day flow variations significantly, and the maximum day was reduced by 5 to 22 percent. The maximum hour was decreased by less than 12 percent.

The use of the washwater recycle systems in place of conventional toilet systems also resulted in significant reductions in daily water use with reductions of 24 and 35 Lpcd relative to a 13.2 L toilet, and 36 and 50 Lpcd relative to an 18.9 L toilet (Table 7). These reductions represented 24 and 22 percent, and 31 and 28 percent reductions in total daily freshwater flow, respectively. Day-to-day variations in freshwater use were not significantly altered. The maximum days were reduced by 7 to 16 percent while the maximum hours were not changed.

Table 7. Reductions in Total Daily Water Use with Innovative Toilet Systems, Lpcd and (%).

Comparison	Extreme Low-Flush Toilets				Washwater Recycle*	
	Microphor [VA]	Monogram [MI]	Ifo [GA]	Ifo [GL]	Aquasaver [NE]	Aquasaver [VAR]
Study System vs Existing**	-	33.6	43.5	-	32.9	22.3
	-	(21.5)	(35.6)	-	20.6	(22.3)
Study System vs 13.2 L	29.1	68.0	15.9	12.4	34.8	24.2
	(27.9)	(35.6)	(16.8)	(22.6)	(21.6)	(23.7)
Study System vs 18.9 L	43.5	102.1	27.2	21.2	50.0	35.6
	(36.6)	(45.4)	(25.7)	(33.4)	(28.3)	(31.3)

* Assumes no freshwater is needed for flushing demand.
** Fixture in use prior to installation of study unit.

The water conservation potentially achievable at typical American households is outlined in Table 8. Daily flow reductions near 30 percent can be expected with these innovative systems.

Table 8. Water Conservation Potential at Typical U.S. Homes with Innovative Toilet Systems*.

Characteristics	Conv. Fixtures		Extreme Low-Flush Toilets				Washwater Recycle**
	Standard	Water-Saver	Micropher	Monogram	Ifo (3L)	Ifo (6L)	Aquasaver
Flush Fol.,L	18.9	13.2	1.9	1.9	3.0	5.3	0
Flush Vol.Red.,%	0	30	90	90	84	72	100
Flow Red.,Lpd	0	79	238	238	222	190	265
Flow Red.,%	0	10.5	31.5	31.5	29.4	25.2	35

*Based on 756 Lpd with 265 Lpd (35%) from 18.9 L toilets at 3.5 npcd.
** Assumes washwater production is sufficient to meet flushing demand.

System Installation

Installation of all three low-flush fixtures can be readily accomplished in a new dwelling. Retrofitting the Microphor toilet may pose minor difficulties in running the required air supply line from the compressed air source to each fixture. Retrofitting the Monogram fixture is readily accomplished. Retrofitting the Ifo fixtures for a conventional U.S. water closet results in a separation of 15 to 23 cm from the rear of the toilet tank to the finished wall, which may be physically or aesthetically unacceptable. Alteration of the drainage system to remedy this situation may be feasible, however.

Installation of the Aquasaver recycle system can be readily accomplished in a new dwelling. Additional plumbing requirements over a conventionally plumbed house include separate toilet supply lines, washwater drain lines, and recycle system vents and connections. Retrofitting such a system is very site specific, and can pose substantial difficulties due to venting requirements, and the need for separate toilet supply lines; especially, in two-story homes with second floor bathroom facilities.

Operation & Maintenance

The operation and maintenance requirements of the low-flush systems were relatively minor, varying from none at all with the Ifo fixtures, to quarterly servicing of the air compressor for the Microphors, and replacement of a broken cutter blade on the Monogram unit with a new improved blade. Power consumption was found to be negligible, varying from 0 with the Ifo units to 0.0016 kWh/use with the Microphors.

The only major operational problem encountered with the recycle system was in maintaining a proper chlorine residual in the recycle water. Residual chlorine values fluctuated widely (0 to 25 mg/L) due to fluctuations in raw washwater quality as well as system design, and caused infrequent odor problems at both homes. Operation at a desired constant residual chlorine level would require considerable monitoring and adjustment. Chlorine use by the system was approximately 0.20 g/L at home NE and 0.30 g/L at home VAR in the form of calcium hypochlorite tablets of 70 percent available chlorine content. Power use by the recycle system was 0.021 and 0.013 kilowatt-hours/cycle at homes NE and VAR, respectively. Scheduled maintenance was performed quarterly in approximately 3 man-hours and consisted of cleaning an influent screen, washing cartridge filters, replenishing chlorine tablets, and removing sludge from the washwater storage tank. Results of this study indicated that the maintenance required for the washwater recycle system may vary from home to home due to differences in water use and raw washwater quality. At home NE, quarterly maintenance as described was readily accomplished and sufficient to maintain the system. In contrast, at home VAR, the influent screen and cartridge filters clogged more severely and made cleaning much more difficult. At this home, more frequent maintenance with annual replacement of the cartridge filters appeared necessary. Sludge accumulations were noted at both homes and consisted of a wispy, black layer occupying the bottom

2 to 10 cm of the storage tank. Analysis of several grab samples of this material revealed high concentrations of organic materials and suspended solids. Sludge disposal via the house sewer system was felt to be the most practical scheme.

User Acceptance

The overall participant assessment of the low-flush fixtures was very positive. The flushing capability of the toilets proved to be satisfactory. For the most part, double flushing was not necessary to clear the bowl and typical cleaning frequencies were sufficient to keep the bowl stain-free. In all study homes, the adult residents indicated they would recommend their type of low-flush toilet to others.

Participant assessment of the recycle systems was both positive and negative. Although the systems provided an adequate water supply for flushing purposes and the use of the recycled water was not generally objectionable for toilet flushing, additional fixture cleaning requirements and occasional septic and chlorine odors reduced the potential for long term user acceptance of these systems.

ECONOMIC ANALYSIS

The costs of the innovative toilet systems as well as two types of conventional toilets are outlined in Table 9. An abbreviated economic analysis was performed for several common residential applications in the U.S. (Table 10). While the data presented are only estimates and subject to considerable variability due to site specific factors, for all of the applications considered the use of the extreme low-flush toilet systems offered potentially significant savings in water supply and sewage disposal costs. Due to the high capital cost of the recycle system, only the holding tank application showed potential cost effectiveness.

Table 9. COSTS OF INNOVATIVE TOILET SYSTEMS* (US$)

Cost Item	Conv. Toilets	Extreme Low-Flush Toilets				Washwater Recycle
		Microphor	Monogram	Ifo (6L)	Ifo (3L)	Aquasaver
Capital	110	550**	500+	365	240	2850
Installation	30	75	30	35	35	400
Operation	0	1/yr	1/yr	0	0	55/yr
Maintenance	?	?	?	?	?	?

*Based upon installation of one fixture in a new dwelling.
**Air compressor and ancillary parts included
+Projected cost of production model

Table 10. ABBREVIATED ECONOMIC ANALYSIS OF INNOVATIVE TOILET SYSTEMS* (US$)

Application/ Assumptions	Conv. Water-Saver	Low-Flush Toilets				Washwater Recycle
		Microphor	Monogram	Ifo (3L)	Ifo (6L)	Aquasaver
Base Conditions						
Installed Cost$	140	625	530	275	400	3250
Incr.Cost		485	390	135	260	3110
over Conv.		86870	86870	81030	69350	96580
Toilet $	0					
Flow Red.,L/yr	28840					
City Water & Sewer						
Water/Sewer Cost Savings @$0.50/ kilolitre,	14.40	43.40	43.40	40.50	34.70	48.30
Payback Pd.,yr						
New-	0	11.2	9.0	3.3	7.5	-**
Retrofit-	9.7	14.4	12.2	6.8	11.5	-**
Rural Holding Tank						
Pumpage Savings @0.5¢/L,$	144	434	434	405	347	483
Payback Pd.,yr.						
New-	0	1.1	0.9	0.3	0.8	7.3·
Retrofit-	1.0	1.4	1.2	0.7	1.2	7.6·
Rural Soil Drain-field-New						
Cost Savings @ $13.50/m^2;90m^2 Req'd with no Flow Reduction$	128	383	383	358	306	425
Net Savings (Loss)	128	(102)	(7)	222	46	(2685)

*Based on the data shown in Tables 8 and 9
** Annual operating costs exceed water/sewer cost savings.
· Pumpage cost savings less annual operating costs.

ACKNOWLEDGEMENTS

The work on which this paper was based, was performed with funds provided by the State of Wisconsin and the U.S. Office of Water Research & Technology under matching grant No.14-34-0001-9103. The generous support of these funding agencies and the efforts of project staff are gratefully acknowledged. A special thank you is given to the residents at each of the study homes for their generous cooperation.

REFERENCES

1. U.S. Water Resources Council: 1978, The Nation's Water Resources - Summary Report. Washington D.C., March.

2. Cohen, S. & H. Wallman: 1974, "Demonstration of Waste Flow Reduction from Households." USEPA Report No. EPA 670/2-74-071.

3. Bennett, E.R. & E.K. Linstedt: 1975, Individual Home Wastewater Characterization & Treatment. Report No. 66, Environ. Resources Center, Colorado State Univ., Boulder, Col.

4. Laak, R.: 1975, Relative Pollution Strengths of Undiluted Waste Materials Discharged in Households and the Dilution Waters Used for Each. Manual of Grey Water Treatment Practice - Part II. Monogram Industries, Inc., Santa Monica, Cal.

5. Siegrist, R.L., M. Witt & W.C. Boyle: 1976, "Characteristics of Rural Household Wastewater." JEED, ASCE, 102, EE3, Paper 12200.

6. Ligman, K., N. Hutzler & W.C. Boyle: 1974, "Household Wastewater Characterizations." JEED, ASCE 100, EEI, Proc. Paper 10372.

7. Siegrist, R.L., W.C. Boyle & D.L. Anderson: 1981, "Field Evaluation of Selected Water Conservation and Wasteflow Reduction Systems for Residential Applications." U.S. Office of Water Res. and Tech. Completion Report. University of Wisconsin-Madison.

8. Standard Methods for the Examination of Water and Wastewater, 14th Edition. American Public Health Assoc., Washington D.C. (1975).

9. Hypes, W.D. & V.G. Collins: 1974, "Reclamation and Reuse of Domestic Wastewater and the Processing of Sewage Solids to Reduce Organics." NASA Langley Research Center, Hampton, Va. Presented at the 1974 Intersociety Conference on Environmental Systems, Seattle, Wash.

10. Siegrist, R.L.: 1978, "Management of Residential Greywater." Proceedings of 2nd Northwest On-Site Wastewater Disposal Short Course, University of Washington, Seattle, March.

11. Donovan, J.F. & J.E. Bates: 1980, "Guidelines for Water Reuse." U.S. Environmental Protection Agency, Report No. EPA 600-8-80-036.

12. National Sanitation Foundation Standard No. 41, Wastewater Recycle/Reuse and Water Conservation Systems. Ann Arbor, Mich. (Nov., 1978).

SESSION III

RESEARCH ON ON-SITE DISPOSAL METHODS IN THE US AND SCANDINAVIA - PAST AND PRESENT

Chairman: R.W. Seabloom

ON-SITE WASTEWATER DISPOSAL RESEARCH IN THE UNITED STATES

James F. Kreissl

Sanitary Engineer/Program Manager
Urban Systems Management Section, Municipal
Environmental Research Laboratory, U.S. Environmental
Protection Agency, Cincinnati, Ohio 45268, USA

During the previous decade concern for the problems of rural America was elevated to a level that the U.S. Congress directed the U.S. Environmental Protection Agency (USEPA) to conduct research to control water pollution in rural areas through the development of new and improved alternatives to conventional gravity sewers and septic tank - soil disposal systems. This concern was due to demographic trends which showed reversal of the historic migration from rural to urban areas and a belated realization that conventional urban sewage collection technologies were inappropriate and uneconomical for rural applications. Also, the limitations of the conventional rural wastewater technology, i.e. the septic tank - soil absorption system (ST-SAS) influenced the lawmakers' charge to the Agency.

In response, the Agency developed and conducted a program of research designed to provide alternative and improved technologies to satisfy the needs of rural residents. These efforts included studies of alternative on-site and collection systems which were both capable of functioning under a variety of conditions which might otherwise preclude proper wastewater disposal and more economical than conventional collection and treatment systems.

A. S. Eikum and R. W. Seabloom (eds.), Alternative Wastewater Treatment, 45–71.

The initial effort was to develop an overall understanding of the state-of-the-art. This included analysis of wastewater characteristics, both in quantity and quality, transmission, treatment, and disposal. These efforts were directed at both on-site treatment and disposal and collection from small groups of homes to entire communities for centralized treatment. The remaining discussion relates only to the on-site treatment and disposal of domestic wastewater aspects of the USEPA research program.

Wastewater is generated from a dwelling in discrete units and varies widely in constituents and in quantity. The average contribution of each major generating fixture group to the total pollutional load has been computed by Siegrist and other major characterization studies and is presented in Table 1 (1). It is important to note the following:

1. The conventional toilet (blackwater) contributes about 80 percent of the nitrogen, 60 percent of the suspended solids (SS). 40 percent of the BOD_5, 30 percent of the phosphorus, and 35 percent of the flow of the combined wastewater, when no garbage disposal is employed.
2. Although the blackwater contains nearly all the microorganisms of the combined waste, the graywater contains numbers sufficient to represent a significant danger to public health.

Characterization of the wastewater represents a basis for the study and design of alternative wastewater systems to overcome specific physical or regulatory constraints at any location. It also permits estimation of the effects of water conservation systems on the discharged wastewater for treatment and disposal system design.

Water conservation is considered to be not only prudent, but often necessary to implement on-site disposal of wastewater. Studies of alternative methods of water conservation in the U.S. have revealed that:

TABLE 1. Average Pollutant Contributions of Major Residential Wastewater Fractions, as Measured in Major Characterization Studies (1), (grams/capita/day)

Fraction	Garbage Disposal	Toilet	Basins, Sinks, Appliances	Approximate Total Contribution
BOD_5	18.0[a] 10.9 - 30.9[b]	16.7 6.9 - 23.6	28.5 24.5 - 38.8	63.2
Suspended Solids	26.5 15.8 - 43.6	27.0 12.5 - 36.5	17.2 10.8 - 22.6	70.7
Nitrogen	0.6 0.2 - 0.9	8.7 4.1 - 16.8	1.9 1.1 - 2.0	11.2
Phosphorus	0.1 0.1 - 0.1	1.2 0.6 - 1.6	2.8 2.2 - 3.4	4.0
Approximate Flow[b] gal/c/d	?	16	29	47
L/c/d	7.6	60.6	109.8	177.9

[a]Mean of Average Values from 6 Major Studies.

[b]Range of Average Values from 6 Major Studies.

1. Passive devices which operate on a batch or predetermined volumetric basis are clearly superior to active devices which are subject to user perceptions. An obvious example is the toilet or water closet. A user will employ this device a certain number (X) of times, generating the flushing volume times X liters per unit of time. Therefore, by reducing X, a known savings will result. A shower restrictor, however, will save an unknown amount from none to the actual flow reduction as a percentage, dependent upon the user's habits and perceptions.
2. Wastewater flow reductions of up to 50 percent are achievable from present average flows of about 175 l/capita/day.
3. Wholesale reductions in water use/wastewater generation are primarily dependent on perceivable water supply shortages and economic penalties for excessive usage in the case of

public supplies. Without these incentives, water conservation has been limited to a relatively minor number of locations.

On-site wastewater disposal research in the U.S. has generally concentrated on disposal systems, although some efforts have been expended on treatment systems prior to disposal of wastewaters. Treatment studies have primarily centered on:

1. The effects of treatment on disposal
2. The performance of alternative treatment devices
3. Modifications to septic tanks.

The influence of treatment efficiency on subsurface soil disposal has been studied by several investigators. There is general agreement that improved effluent quality can result in less-intensive clogging of coarser, unstructured soils, such as sands. However, advocates of higher treatment efficiency maintain that these effects are also evident with finer, more structured soils. The USEPA-sponsored studies at the University of Wisconsin failed to determine any significant influence on effluent quality within achievable ranges on slowly permeable soils (2).

Theoretically, an aerobic system should perform more efficiently than a septic tank. However, several field studies have failed to demonstrate consistent performance of these systems. The difference between performance under controlled, idealized testing and field studies lies in the lack of regular operation and maintenance (O&M), the variability of wastewater, and the physical environment in which the unit operates. Table 2 displays the results of several field studies (3). Given the variability of the data shown in the table, any attempt to reduce the size of the soil absorption system based on the superior quality of aerobic units would appear to be short-sighted at best.

TABLE 2. Aerobic Unit Effluent Qualities as Determined by Various Investigations (3), in mg/l.

Source Location/Year	BOD_5 Mean	BOD_5 Range	Suspended Solids Mean	Suspended Solids Range
Wisconsin, 1978	37(112)[a]	0 - 208	39(117)	3 - 252
Canada, 1967	47(86)	10 - 280	94(74)	18 - 692
New York, 1974	92(146)	-	94(146)	-
Colorado, 1975	144(393)	10 - 824	122(251)	17 - 768
Maryland, 1975	36(124)	3 - 170	57(132)	4 - 366
Kentucky, 1977	37(167)	1 - 235	62(167)	1 - 510

a) Number in Parenthesis is the Number of Samples.

Considering that increased capital and O&M costs of aerobic units greatly exceed any potential savings in reduced SAS sizing, there is little economic reason for the substitution.

Several investigators have looked at the anaerobic upflow filter as a means of improving septic tank performance. Because of the idealized hydraulics of these devices they are capable of improving the effluent quality of septic tank effluents. Questions related to increased O&M required for anaerobic filter cleaning and increased sulfide concentration in the effluent have yet to be addressed by long-term field studies. Also, the cost-benefit ratio remains in question for those systems discharging to finer, more structured soils for absorption.

The questions of compartmentalized tanks and improved inlet and outlet structures continue to be discussed without sufficient data for resolution. With rare exception the multiple-compartment septic tank appears to perform better than the single-compartment design of equal volume (4,5,6). Unfortunately, the code writers of most U.S. jurisdictions have not been convinced that improved tank design and performance are an adequate tradeoff for slightly increased costs of fabrication and maintenance. Similarly, improved outlet arrangements have not been implemented in the field despite som evidence that improved effluent quality results from their use. Several authors have proposed design improvements (4,5,7).

Disposal methods for on-site systems are divided into three major categories, subsurface, surface and atmospheric disposal, although many systems employ more than one major method. The conventional ST-SAS is designed for subsurface disposal of wastewater, and remains the most desirable system in use today where site conditions are favorable. This system, depicted in Figure 1, has been misapplied extensively over the past several decades, often resulting in high failure rates. Such misapplication has been due to a variety of factors, including poor site evaluation and design, inadequate construction methods, and the lack of adequate alternative systems other than conventional gravity sewers. Despite the design and construction shortcomings extant with the use of these systems, studies have demonstrated system service half-lives of 50 years or more.

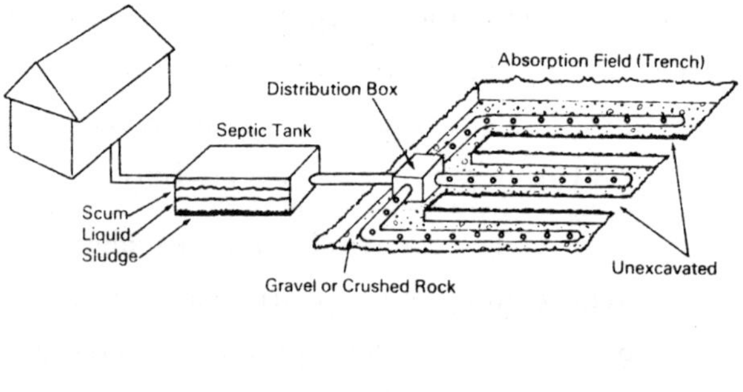

Figure 1. Conventional septic tank-soil absorption system.

The conventional ST-SAS is limited to locations with moderately permeable soils and relatively deep water tables or impermeable strata. Also, the definition of what constitutes a conventional soil absorption system varies between locations. For example, in many locations absorption beds and pits are common, while trenches abound elsewhere. The design of trenches varies as well. Deep, narrow trenches are fairly common in some areas, while wider, shallower trenches are common in others. Therefore, as a basis of comparison in this discussion "conventional ST-SAS" is defined as a septic tank followed by gravity loading of the subsurface soil absorption system through 10 cm (4 in) pipes with multiple openings throughout the SAS length.

Since 1973 the USEPA Small Flows Research Program has attempted to support both improved understanding of conventional on-site systems and development of viable alternative systems to overcome specific site limitations which preclude the use of conventional systems. Significant progress has been made in these research and demonstration efforts, with the result that several areas of the U.S. have adopted these technologies to practice. The following discussion provides some of the available results obtained from field installations of these systems and some pertinent research results.

PRESSURE DISTRIBUTION/DOSING SYSTEMS

The concept of uniform distribution of pretreated effluent to the disposal trench has been employed in an attempt to offset the problem of "creeping progressive failure" and the potential groundwater contamination which accompanies the development of this phenomenon. Because of the size of the conventional distribution pipe (10 cm) and the small and discrete nature of household discharges, most of the wastewater is applied to a very small portion of the trench near the inlet. To reduce the rate of clogging and to eliminate localized overloading (see Figure 2)

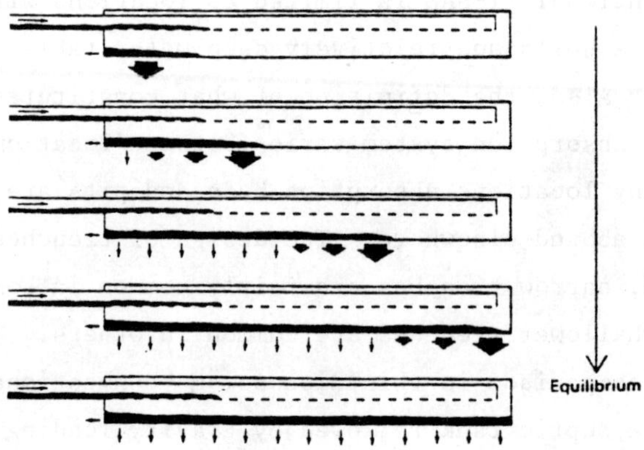

Figure 2. Hydraulic characteristics of a conventional gravity-fed soil absorption system after initiation of operation.

pressure distribution systems have been employed. University of Wisconsin studies (2) employed dosing rates of 3 times normal levels to a system in a silt loam soil, resulting in negligable clogging in 4 years of operation. Carlisle (8) has reported superior results to conventional systems with a pressure distribution system design in North Carolina, noting only two problem systems in 33 installed. Ronayne (9) reports the successful use of pressure distribution systems where high groundwaters and coarse soils exist. In a coarse (gravelly sand) soil, design loadings are 1.5 cm/day and measurements 90 cm below the soil revealed fecal coliform counts of 10/100 ml and about 50 percent nitrogen removal when dosed at 8 to 10 times per day (10). In essence the data support reduction of the vertical separation requirement from 1.3 m to 0.8 m when pressure distribution is employed (9). Jones (11) has discussed use of timer controls instead of the normal level controls preset to dose specific volumes. The use of pressure distribution is most favorable with coarser soils, but is also applicable to structured soils with abundant macropores to minimize contamination of groundwater. Field tests in slowly permeable soils showed pressure distribution to perform similarly to conventional gravity dosing in soils without significant macropores in Oregon (9). The sand filter part of this study, reported later in this discussion, clearly verified the work of Bouma (12) in delineating the advantages of small, uniformly applied doses in reducing the number of mirco-organisms penetrating through sand or finer soils with significant macropores.

ALTERNATING BED SYSTEMS

The concept of restoring the original permeability of soil absorption systems by allowing them to rest has been demonstrated in the laboratory and several field sites. Documentation is, however, sparse. Wiegand (13) reported three successful systems in West Virginia, but their large size (153 m^2 of bottom area) would appear to limit their applicability. Since 1973, Fairfax County, Virginia, has required that an absorption bed be split into two sections (halved) with annual diversion between halves. No failures of these systems have been reported by county officials since this regulation has been in effect. University of Wisconsin laboratory studies (2) determined that only 3 to 4 weeks of rest may be required to restore failed sandy systems to acceptable permeability.

Further studies to determine the independent variables for adequate resting periods for permeability restoration in different soils are underway by the University of Wisconsin and Pennsylvania State University. The alternating bed concept is illustrated in Figure 3.

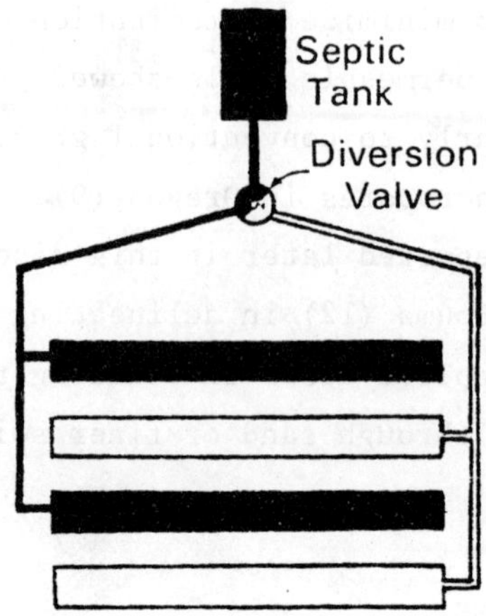

Figure 3. Schematic diagram of an alternating bed system with its diversion value.

MOUNDED SYSTEMS

In areas with marginal soils, high groundwater, or insufficient soil depth which preclude conventional systems, one alternative is the mound system. In essence the mound is designed to provide treatment and distribution of septic tank effluent prior to its introduction into the natural soils. This concept was originally employed in North Dakota, but it has been refined by several others, most notably, Bouma at the University of Wisconsin. Although Witz (14) has reported on 25 years of experience with the raised and primarily gravel-filled Nodak system, no survival data were included. Wiegand (13) reported a survey on 12 Nodak systems in Wood County, West Virginia, noting that 7 were failing, primarily due to poor installation and siting of the system. He also described a modified (West Virginia) Nodak which employed pressure distribution and additional sand. The failure rate for that system was 40 percent of the 38 installations surveyed, due reportedly to poor installation and maintenance.

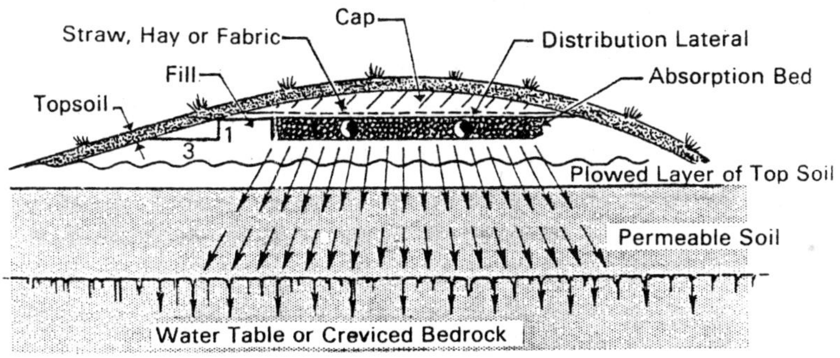

Figure 4. Cross-section of a Wisconsin mound used for locations with inadequate unsaturated (permeable) conditions.

The Wisconsin mound, depicted in Figure 4, employs pressure distribution and sand as a fill material. Its record of success in a recent study of 640 installations in Wisconsin was 99.5 percent, with 44 percent nitrogen removal (15). Other investigators reporting on their experience with Wisconsin mounds have noted similar successes. Wiegand noted that one of five surveyed Wisconsin mounds failed due to poor installation, while Carlisle (8) and Seabloom et al. (16) reported no failures in three and two installations, respectively. Paeth (10) described generally excellent performance in Oregon, except when groundwater levels rose to the surface after heavy rains, which is to be expected with any subsurface system under these circumstances. Minnesota (17) has enjoyed similar success with the mound system. Several variations in mound design have been employed with limited success. The New Mexico version has employed plastic liners and the Pennsylvania version clay diking, both without pressure distribution. A recent survey of the latter resulted in a 43 percent failure rate of the 81 installations observed (18). Similarly, several failures have been recounted in New Mexico, with resulting recommendations to eliminate these design deficiencies (19).

Oversized mounds employing gravelly loam fill have had an 89 percent success rate in Erie County, New York (20). Conversely, undersized mound design employing aerobic pretreatment have been observed to enjoy only a 67 percent success rate in Maryland (21). The Wisconsin mound appears to offer the best chance of success, but is considered somewhat costly and requires knowledgeable designers, inspectors and installers.

ARTIFICIALLY DRAINED SYSTEMS

In areas where permeable, drainable soils exist, but are unsuitable for conventional systems due to high groundwater levels, artificial drainage may permit site utilization by lowering the water table (see Figure 5). Sites must have sufficient relief to permit free discharge of drained groundwater. Ronayne (9) and Paeth (10) have reported on 12 systems utilizing 1.2 and 1.8 m deep interceptor trenches surrounding 0.6 m deep absorption fields with success. The absorption and drainage lines were 3 meters apart for the 1.2, and 6 meters apart for the 1.8 m deep drains, respectively. Measurements of groundwater levels below the SAS trench bottom were always greater than 0.2 m, while without the drains they would have been in the trenches. Fecal coliform counts have indicated no significant increase in the drained groundwater, although nitrate levels were higher than background. This technique has often been misapplied by attempting to dewater poorly drainable soils. Severe failure conditions have occurred under these circumstances.

SHALLOW TRENCH SYSTEM

In areas where groundwater levels or impervious layers occur too close to the ground surface to permit conventional on-site systems, shallow systems have been employed. Both gravity and pressure distribution have been used and often some

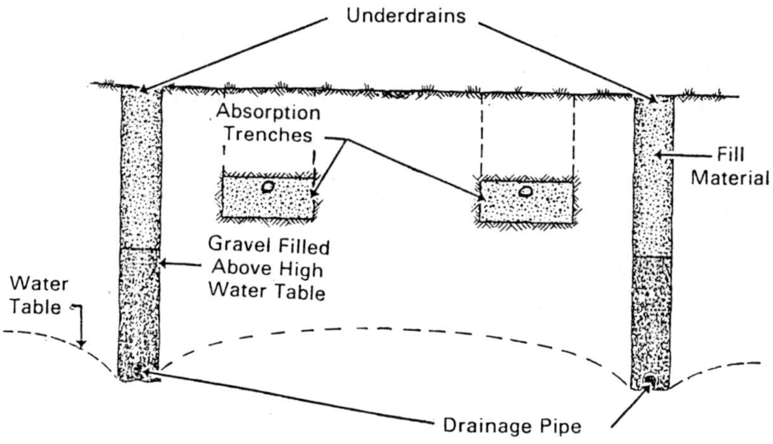

Figure 5. Underdrained soil absorption system.

thickness of fill or capping is applied over the natural surface to protect it from freezing. A typical system is shown in Figure 6. Wiegand (13) reported no failures with 4 gravity-fed systems employing 0.3 to 0.45 m deep trenches with 0.15 to 0.3 m of capping. Similar pressure-distribution systems have also performed well in Oregon (9), and have been used in Minnesota with no performance data available. Bouma (12) has suggested that shallow systems are often desirable since the shallow depth maximizes the benefits of evapotranspiration (ET), takes advantage of the usually more permeable surface layers of the soil, maximizes the unsaturated soil zone depth and its associated treatment capacity, and in clayey soils, avoids the deeper, usually wetter, strata where smearing of surfaces during construction is most likely.

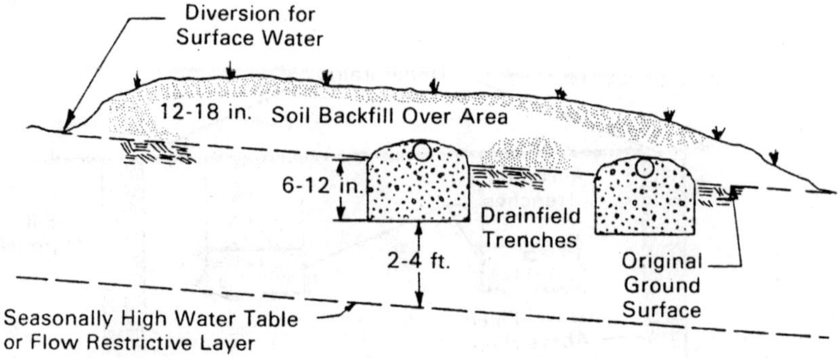

Figure 6. Trench system employed to overcome marginally inadequate unsaturated zone conditions.

EVAPOTRANSPIRATION SYSTEMS

In certain climatic conditions where conventional on-site systems are not administratively or technically feasible, evapotranspiration (ET) systems have been employed successfully. Next to the conventional ST-SAS this system has been the most misapplied of all site alternatives. Because early research (22) was incomplete, a misconception developed that economic, pure ET systems performed well in cold, wet climates. Thermodynamic studies of the ET (23) have disproven this assumption along with numerous failures of field installations. Another misconception disproven thermodynamically was the beneficial effect of aerobic effluent in the ET process. Although numerous climatic misapplications have failed, several have succeeded due to the fortuitous occurrence of holes in the thin plastic liners. Properly, an ET system requires a plastic liner of at least 0.25 mm (10 mil.) thickness along with careful placement between sand layers to ensure its integrity. A generic ET bed system is shown in Figure 7, but trench systems are also used. Wiegand (13) reported a successful installation with no other information,

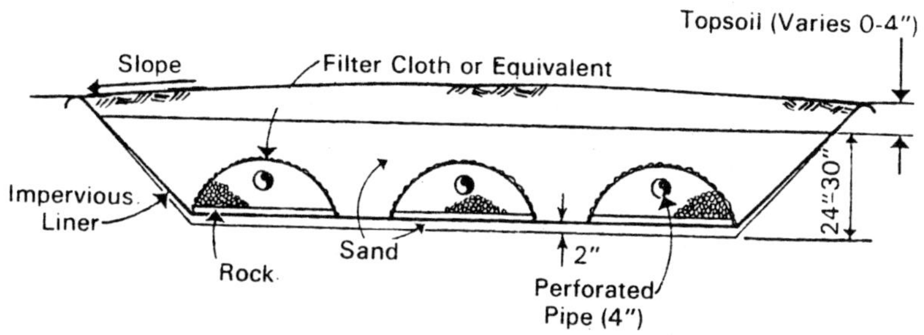

Figure 7. Typical ET bed cross-section.

while 13 of 16 undersized commercial ET installations in an arid region of Oregon displayed leakage through the 0.1 mm (4 mil.) liner and two of the other three overflowed (9). A 700 m^2 system of non-commercial design, also in Oregon, has functioned well (9). Several systems have failed in New Mexico, despite its arid climate, due to poor construction and design practices, notably the use of sand which was too coarse to properly "wick" the wastewater to the surface (19). Several failures have not unexpectedly been observed in Kentucky, Vermont, Maryland, Pennsylvania and other eastern U.S. locations. The most extensive study of ET was done at the University of Colorado (24). The study optimized sand sizing (d_{50} = 0.1 mm), tested aerobic and septic effluent to verify the thermodynamic study cited above, and used multiple lysimeter testing to determine the effects of solar radiation, wind, vegetative covers, soil covers, and effluent levels on ET, in addition to determining the optimum design loading for Boulder, Colorado, i.e., 1.2 mm per day (0.03 gpd/sf). About 2/3 of the evaporation experienced was due to unsaturated air movement, with the other 1/3 due to solar radiation. Native soil covers were inhibiting to ET, and ET decreased with increased depth of effluent. Vegetation did not appear to offer assistance on a year around basis, as increased ET during the growing season appeared to be offset by reduced winter rates. However, vegetation is highly desirable for summer dwellings where use coincides with high transpiration. Such seasonal systems have been reported to be successful in Wyoming and New York (25,26).

EVAPOTRANSPIRATION-ABSORPTION SYSTEMS

The evapotranspiration-absorption (ETA) system is generally the same as the ET system in Figure 7, except no liner is used in order to obtain as much infiltration as possible. In actuality, such systems are shallow absorption systems, designed to get the maximum benefit from ET. Ronayne (9) and Paeth (10) reported the successful use of 0.6 m deep serial distribution trenches and beds in Oregon with horizontal areas of from 170 to 250 m^2. However, because of the excess size of these beds and the lack of wicking soil, the researchers believe that disposal was by absorption alone. Several ETA systems have been reported in Ohio (27), but results confirmed that the undersized trench systems worked well only in areas where soils were permeable and hydraulic loading was low. Also, the requirement for aerobic units and the use of inappropriate tree species, i.e. Scotch Pines (Pinus sylvestris), caused economic hardship on several users.

SAND FILTERS

Often, site conditions and/or regulations preclude any subsurface or ET solution for wastewater disposal. In some locations in the U.S. surface discharge is permitted. For many years the buried sand filter has been used with some success, but several problems have resulted from improper construction, design and materials. Any such deviations from the requirements of proper filter design and operation become economic hardships due to the inaccessability of buried filters.

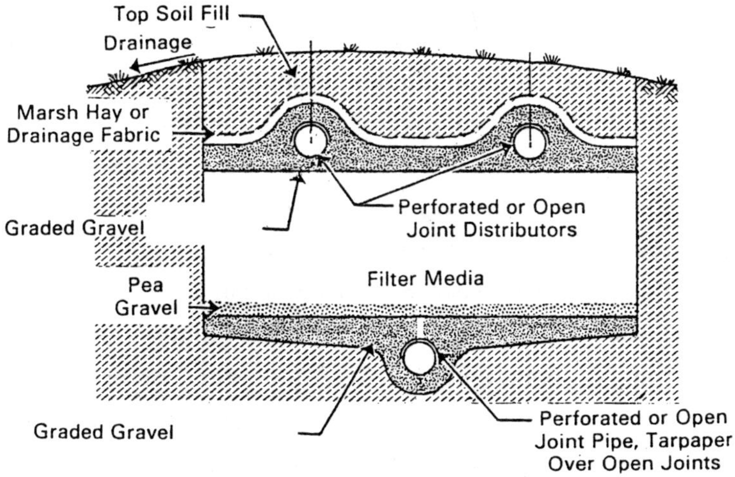

Figure 8. Typical buried sand filter.

Buried sand filters, as shown in Figure 8, require 0.6 to 0.9 m of relatively uniform (uniformity coefficient (UC) $\leq$ 3.5) sand with an effective size (ES) of 0.5 to 1.0 mm dosed at a design rate of no more than 4 cm/day, split into at least 2 doses per day. Field data indicate that the BOD_5 and SS of effluents from these systems are generally less than 10 mg/1 (28). Properly designed systems have been in service for many years.

Because of concerns over the size and cost of these filters and the expense of repairs, several alternative designs have been studied in recent years. Ronayne (29) has reported on four years of experience with modified filters in Oregon which employ uniform (UC $\leq$ 3), fine (ES = 0.3 to 0.4 mm) sand of 0.6 m depth, pressure dosing to ensure uniform loading, and a hydraulic loading of 5 cm/day at peak flow. Effluent concentrations of BOD_5 and SS were 3 and 7 mg/1, respectively, with nitrogen removals of 40 percent. By decreasing the volume of the dose from 2.4 cm/dose to 1 cm/dose effluent fecal coliform, BOD_5 and SS concentrations were reduced from 1120/100 ml, 4.8 mg/1, and 3.3 mg/1 to 111/100 ml, 1.9 mg/1, and 1.6 mg/1, respectively.

University of Wisconsin studies (2) of higher rate intermittent filters were designed to reduce the areal requirements and high initial cost of buried sand filters, by accepting additional O&M costs due to periodic sand surface restoration. This type of filter employs a free-access design with similar sand size (ES = 0.35 to 1.0 mm) and uniformity (UC $\leq$ 3.5), and a design loading rate of 8 to 20 cm/day for septic tank effluent. Depth of sand is the same as with the buried design. Two filters are used when filtering septic tank effluent to permit resting of previously clogged sand (28). Two years of operation with level-controlled dosing of 5 cm/dose yielded effluent concentrations of BOD_5 and SS of 10 mg/l and 14 mg/l, respectively, while total nitrogen was unchanged. Phosphorus removals reduced from an initial 20 to 30 percent to none as the filter runs progressed. Filter runs at 20 cm/day loading were approximately 90 days under these conditions for the 0.45 ES sand (2). It is interesting to note that no nitrogen removal occurred at the 20 cm/day loading while 40 percent nitrogen removal has been reported at 4 cm/day in mounds and filters.

In the States of Illinois and Oregon several installations of another filter modification have been monitored. This system is called a "recirculating filter", and employs recirculation to obtain a high degree of treatment without the odor potential of intermittent systems. As shown in Figure 9, recirculating filters require an additional tank which receives both septic tank effluent and filtered effluent for discharge to the filter in doses of 5 cm depth as often as 48 times per day. The recycle ratio is generally larger than for intermittent filters, usually in excess of ES = 1.0 mm (although smaller sizes have been employed) to accommodate a forward flow-loading of 12 to 16 cm/day at a 5:1 recirculation ratio. One of the first of about 1,000 Illinois installations averaged 4 mg/l of BOD_5 and 5 mg/l of SS in effluent samples (28). Four similar systems in Douglas

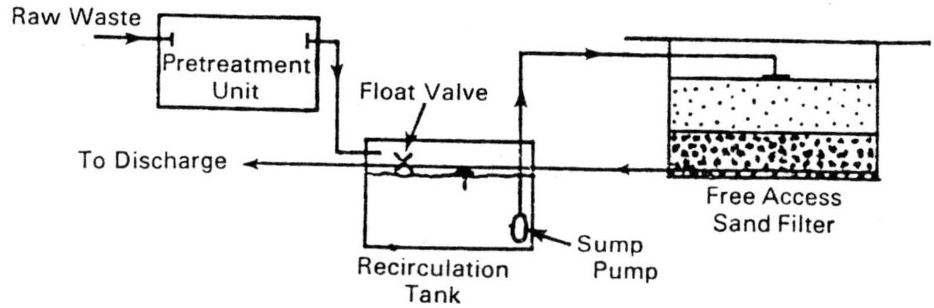

Figure 9. Schematic view of a recirculating sand filter.

County, Oregon, with larger, more uniform, sand (ES = 1.2 mm; UC = 2) produced average BOD_5 and SS effluent concentrations of 4 mg/l and 3 mg/l, respectively, while fecal coliform levels averaged about 10^5 per 100 ml (38). At the same wastewater design loading, i.e. 12.5 cm/day at peak flow (area = 13.4 m^2), Ronayne (29) reported that four more Oregon recirculating filters produced BOD_5 and SS effluent concentrations of 2 mg/l and 1.5 mg/l, respectively, with about 50 percent nitrogen removal and fecal coliform means of 15 to 440 per 100 ml.

Other filter designs have been employed to a lesser extent. In any case, these systems must produce reliable, consistent high-quality effluents with minimal O&M requirements.

BLACKWATER/GRAYWATER SYSTEMS

In recent years much interest has been expressed in wastewater separation and recycling techniques. Black (toilet) wastes are usually handled without water carriage, primarily by composting or incinerating toilets. Although other non-water systems have been employed for in-house usage, the practicality is yet to be proven. Although thousands of composting toilets have been installed in the U.S. in the last ten years, until recently very little data were available, except for some Norwegian studies. Ronayne (9) reported on several Oregon installations over the

past four years. In Oregon, observed experience by owners indicated that more than 2/3 of the units encountered fly problems and a similar number encountered excess liquid buildup. The former problem was associated with large and small units, while the latter was primarily a small unit problem. More than 1/2 of all the small units also suffered excess waste accumulation problems. Further owner response indicated that pesticides were the most common method of eliminating insect pests, primarily the vapor-emitting solids. Approximately 2/3 of the owners reported noticeable to disagreeable odors from their composting toilets.
A NBS report (21) recounts several visits to composting toilet installations in the eastern U.S., noting that most of the larger units operated acceptably for the users, while most of the smaller units were failing in terms of odors and waste accumulation.
Recent studies in California (30) illustrate that all large slanted-flow commercial units were soggy (moisture 67 percent), pile temperatures averaged 18 ^{o}C, over 80 percent had fly and liquid buildup problems, and about 1/3 had noticeable odor and structural failure problems. Additional problems included short-circuiting of wastes and periodic isolation of parasitic and bacterial pathogens in the final product chamber. Investigators noted the inadequate use of bulking agents, the need for physical mixing of the pile, and the ubiquitous presence of coliforms (10^{2} to 10^{10} per gram of solids) in the final chamber. This study also involved two types of small commercial composting toilets. Most temperatures were lower than 40 ^{o}C (1/2 below 25 ^{o}C), all but one exhibited unacceptable odors, little decomposition of wastes occurred, 3 of 8 exhibited fly problems, 4 of 8 had liquid buildup, and structural problems were common with topping bars and rotor handles. Hookworm ova and larvae, ciliates, and _Yersinia_, _Citrobacter_, _Enterobacter_, _Klebsiella_, _Providencia_, and _Hafnia_ bacterial species were isolated from the liquid buildup samples. Vector studies were performed to quantify various types of organisms found. Best results in terms of reduced

numbers of flying insects and mites were with the small unit supplied with a rotor, but all units were positive for their presence. Fly problems were found to be most controllable by proper sealing and screening techniques, immediate addition of kitchen scraps and by good sanitation habits. In general agreement with other researchers, the most desirable method of final product disposal was found to be burial in an area where no contact with food crops is likely. The authors stressed the need for third-party management of these systems and improved user education programs.

Incinerating toilets employing gas, electricity or a combination have also been used in several areas of the U.S., but not frequently in year round homes. Of six home installations in Kentucky during the early 1970's, five had been abandoned by 1978, and the other house was unoccupied. Incomplete oxidation, odors, high operating costs, and frequent repairs were cited as reasons for failure (31). NBS (21) reported high operating costs and odors with some units, but some success in that 2 of 4 units were acceptable to the users.

Several other waterless toilet designs, including oil-carriage and various types of chemical and biological systems have been promoted to take advantage of the fact that 1/3 of the water usage and most of the nitrogen waste contribution are toilet associated. However, since it is relatively rare that nitrogen loadings are limiting, the most obvious need has been to conserve water. In such cases, the reduced-flush toilet has generated more interest in recent years. Although there is little U.S. data on these systems, the reduction of toilet flushing volumes from the usual 16 liters to relatively low levels is considered a direct savings in overall conservation. One commercial unit which is relatively popular, employs air pressure and requires only 2 liters per flush. NBS (21) has observed several of these units which generally operated well. Siegrist (32) has studied this unit and two others which were also quite acceptable to users. Low power consumption (6 to 10 kWh/yr), low maintenance and similarity to conventional toilets influenced user reaction. The daily blackwater volume reductions achieved varied from 87 percent for the air-assisted unit to 54 to 77 percent for the low-flush gravity units from Sweden.

The concept of wastewater separation, although primarily related to waterless toilets, has resulted in several studies of graywater, since it has been well established that this wastewater is not innocuous and requires similar treatment and disposal to combined wastewater. Graywater characterization and treatment studies at the University of Wisconsin and in Oregon have shown that septic tank treatment results in an effluent of great similarity to the combined wastewater, with the lone exception of lower nitrogen content (9,33). Specific pathogen studies on graywater have isolated Pseudomonas aeruginosa, Enterobacter, Citrobacter, Klebsiella pneumoniae, and various ciliates, flagellates, larvae, and mite-eggs (1, 30). No virus isolations were successful (30). In light of the above, no direct recycle of graywater with the home is prudent. In any graywater treatment sequence the initial step should provide for grease removal. The conventional septic tank is normally used, but some smaller tanks are commercially available. Two studies (9,30) reported generally poor treatment by a 0.75 m^3 (200 gallon) commercial graywater septic tank. Various commercial graywater systems are available including pea gravel filters and various settling/filtration combinations. None of these units have been found to significantly improve graywater quality to meet any surface discharge standard (9,30,33). A recirculating sand filter was used for graywater treatment in Oregon, producing an effluent (BOD_5 = 2.2 mg/l; SS = 10.5 mg/l) similar to a combined wastewater system of this design (9). At this time it appears that no data are available to justify any reduction in the treatment/disposal of graywater from that required for combined wastewater systems other than the flow reduction factor of one-third.

OTHER TREATMENT/DISPOSAL SYSTEMS

Although several other design modifications have been attempted in the U.S. in recent years, few have generated sufficient interest, in terms of passing an engineering evaluation of their potential practicality and cost-effectiveness, to be studied in greater detail. Although isolated local conditions may make some of these alternatives feasible, their priority for national interest has been considered low. Among these systems, complete and partial recycle probably represents the greatest potential in water-short areas, but often intelligent engineering will provide more cost-effective solutions.

ANCILLARY STUDIES

The most significant ancillary research resources have been devoted to residuals handling, management, and facility planning to improve implementation of improved on-site technologies. The most comprehensive of these efforts has been devoted to septic tank pumpings (septage) treatment and disposal. Over the past seven years a variety of research studies have investigated several alternative techniques at pilot- and field-scale. Presently, these data are being analyzed and combined into a handbook of septage handling techniques in cooperation with the Norwegian Institute for Water Research (NIVA) to provide engineers with a single, comprehensive source of design operation and selection information.

As data became available on several of the alternative on-site and small community systems described earlier, it became apparent that some form of centralized management would be necessary to realize the potential technical and economic benefits of these technologies. Therefore, a significant level of research funds has been expended to define various management alternatives already in use in a few areas of the country and to develop a user's guide to assist local officials and engineers to choose the optimum management strategy for their local administrative and legal constraints and for the alternative technologies to be incorporated in their area. This user's guide is to be released in early 1982.

Also, the need to provide assistance to engineers in evaluating small community problems was evident. Since most U.S. engineers have been educated in a system which concentrates only on conventional urban technologies, their unfamiliarity with unique rural problems resulted in their ignorance of the inappropriate nature of their known technologies. Therefore, some recent research efforts have been devoted to providing the engineering community with new approaches to problem identification and solution development for small communities.

One additional aspect worthy of note as a significant research concern has been that of proper construction techniques. Studies now completed at the University of Wisconsin for EPA have attempted to quantify the soil damage which can occur during construction of subsurface systems (34). In essence damage to fine loamy soils was greatest during wet conditions, but during dry conditions more prominently structured loams were damaged less than poorly structured loams. Other interesting facts include the relationships between pore size and damage and relative damage due to equipment and hand digging. The results of this work will be available in the early 1982.

SUMMARY

The alternative on-site technologies described herein have been developed over the past several years, often with the assistance of the USEPA Small Flows Research Program. These alternatives are being implemented at certain locations in the U.S., and their performance will be the primary determinant of the success or failure of the research activities of the past decade. As experience with these alternatives becomes greater, it is projected that success rates will improve, costs will reduce, and new plateaus of technical expertise will be attained. This scenario will depend heavily on how well the EPA research program performed its function.

REFERENCES

1. Siegrist, R.L.: 1978, "Waste Segregation to Facilitate Onsite Wastewater Disposal Alternatives." Proceedings of the Second National Home Sewage Treatment Symposium, ASAE Pub. No. 5-77.

2. Small Scale Waste Management Project: 1978, Management of Small Wastewater Flows. USEPA Report No. 600/2-78-173.

3. Hutzler, N.J., L.E. Waldorf & J. Fancy: 1978, "Performance of Aerobic Treatment Units." Proceedings of the Second National Home Sewage Treatment Symposium, ASAE Pub. No. 5-77.

4. Weibel, S.R., C.P. Straub & J.R. Thoman: 1949, Studies On Household Sewage Disposal Systems - Part I, Federal Security Agency, Public Health Service Publication.

5. Winneberger, J.T.: 1980, Septic Tank Systems, One Consultant's Toolkit, Hancor, Inc.

6. Laak, R.: 1980, "Multichamber Septic Tanks." Journal ASCE-EED, Vol. 106, No. 3, p. 539.

7. Baumann, E.R., E.E. Jones, W.M. Jakubowski & M.C. Nottingham: 1978, "Septic Tanks." Proceedings of the Second National Home Sewage Treatment Symposium, ASAE Pub. No. 5-77.

8. Carlisle, B.L.: 1979, "Development of Alternative On-Site Septic Waste Disposal Systems for the Coastal Zone of North Carolina." In: Fosse Septique, N.C. Division of Health Services Publication No. 1.

9. Ronayne, M.P.: Oregon On-Site Experimental Systems Program, USEPA Grant No. S806349, unpublished progress reports.

10. Path, R.C.: 1980, "Oregon's Experimental Program." Proceedings of 3rd Northwest On-Site Wastewater Disposal Short Course.

11. Jones, E.E., Jr.: 1980, "Improved Controls for Septic Tank Effluent - Soil Treatment Systems." Paper presented to ASAE North Atlantic Region Meeting, ASAE Paper No. NA80-208.

12. Bouma, J.: 1977, Soil Survey and the Study of Water in Unsaturated Soil, The Netherlands Soil Survey Institute, Soil Survey Paper No. 13.

13. Wiegand, R.G.: 1979, "Performance of Alternative On-Site Sewage Systems in Wood County, West Virginia." Jour. of Env. Health, 42, 3, p. 133.

14. Witz, R.L.: 1975, "Twenty-Five Years with the NODAK Waste Disposal System." Proceedings of the National Home Sewage Disposal Symposium, ASAE Pub. No. PROC-175.

15. Harkin, J.M., C.J. Fitzgerald, C.P. Duffy & D.G. Kroll: 1979, Evaluation of Mound Systems for Purification of Septic Tank Effluent, U.S. Department of Interior, OWRT Report No. WIS WRC 79-05.

16. Seabloom, R.W., D.A. Carlson & J. Engeset: 1981: Individual Sewage Disposal Systems. Report to DHUD and Was. DSHS by University of Washington.

17. Hansel, M.J. & R.E. Machmeier: 1980, "On-Site Wastewater Treatment on Problem Soils." Jour. WPCF, 52, 3, p. 548.

18. Petersen, G.W. & D.D. Fritton: 1979, Evaluation of Mound Systems for Renovation of Septic Tank Effluent, U.S. Department of Interior, OWRT Report by Pennsylvania State University.

19. Metric Corp.: 1981, Evaluation of Evapotranspiration and Mound-Type Liquid Waste Disposal Systems - Los Lunas - Belen Area, New Mexico, Report to New Mexico Env. Imp. Div.

20. Barry, D.E. & R.M. Donnelly: 1979, "Experiences with Mound Systems in Erie County, New York." Individual Onsite Wastewater Systems, Ann Arbor Science.

21. Winter, F.: 1981 Onsite Wastewater Systems - Current Practices and a Proposed Basis for Evaluation, U.S. Department of Commerce, NBS Report No. NBS1R 81-2210.

22. Bernhart, A.P.: 1973, Treatment and Disposal of Wastewater from Homes by Soil Infiltration and Evapotranspiration, Vol. 1, University of Toronto Press.

23. Tanner, C.B. & J. Bouma: 1975, "Influence of Climate on Subsurface Disposal of Sewage Effluent."Proceedings of 2nd National Conference on Individual Onsite Wastewater Systems, National Sanitation Foundation, Ann Arbor.

24. Bennett, E.R. & K.D. Linstedt: 1978, Sewage Disposal by Evaporation - Transpiration, USEPA Report No. 600/2-78-163.

25. Hasfurther, V.R., D.H. Foster & G.D. Lloyd: 1978, "Sizing an Evapotranspiration Waste Disposal System for Summer Operation." Proceedings of 2nd National Home Sewage Treatment Symposium, ASAE Pub. No. 5-77.

26. Salvato, J.A., Jr.: 1975, "Problems and Solutions of Onlot Sewage Disposal." Proceedings of 2nd National Conference on Individual Onsite Wastewater Systems, National Sanitation Foundation, Ann Arbor.

27. Mid-Ohio Regional Planning Commission: 1979, An Inventory and Analysis of Existing ETA Systems in Franklin County, Ohio. Draft Report to Ohio EPA.

28. 1980, Design manual - Onsite Wastewater Treatment and Disposal Systems, USEPA OWPO/MERL/TT Publication.

29. Ronayne, M.P.: 1980 "Sand Filters - Oregon Experience." Proceedings of 3rd Northwest On-Site Wastewater Disposal Short Course.

30. Enferadi, K.M. & J.H. Poorbaugh: 1981, A Field Investigation of Several Black Waste Composting Systems, Draft Report for EPA Grant No. R805942.

31. Abney, J.L.: 1980, Evaluation of 19 On-Site Waste Treatment Systems in South-eastern Kentucky, USEPA Report No. 600/2-80-101.

32. Siegrist, R.L.: 1980, "Residential Waste Flow Reduction with Low-Flush Toilet Fixtures." Proceedings of 3rd Northwest On-Site Wastewater Disposal Short Course.

33. Siegrist, R.L.: 1980, "Greywater Treatment by Coarse Media Filtration of Septic Tank Effluent." Proceedings of 3rd Northwest On-Site Wastewater Disposal Short Course.

34. Small Scale Waste Management Project: 1981, Design and Management of Subsurface Soil Absorption Systems, Draft Report for USEPA Grant No. R805531.

ON-SITE WASTE DISPOSAL [illegible] UNITS [illegible]

1. [illegible] 1978. [illegible] Sept. Drains [illegible] Treatment "[illegible] of [illegible] Onsite Wastewater Systems [illegible] Research [illegible]

2. [illegible] 1978. Sewage Disposal [illegible] Report No. [illegible]

3. [illegible] 1976. "[illegible] for on-site Waste Treatment System [illegible] Operation." Proceedings of the National Home Sewage Treatment Symposium, ASAE Pub. [illegible]

4. [illegible] 1975. "Problems and Solutions of Onsite Sewage Disposal." Proceedings of the National Conference on Individual Onsite Wastewater Systems, National Sanitation Foundation, Ann Arbor [illegible]

5. [illegible] 197[illegible] [illegible]

6. [illegible]

7. [illegible]

8. [illegible]

RESEARCH ON ON-SITE DISPOSAL METHODS IN NORWAY - PAST AND PRESENT

Rolv Kristiansen

Research Coordinator, Agricultural Research Council
of Norway, Section Soil Pollution Research,
1432 Ås - NLH, Norway

In 1970, the research on on-site disposal methods was initiated as one of the sub-projects of an extensive national research program which comprised a variety of research areas dealing with wastewater treatment (Liseth, 1980). The Agricultural University of Norway (NLH) was given responsibility for investigating soil as a recipient and renovating medium for sludge and wastewater, including testing the performance of biological toilets. Approximately NOK 7.05 million were allocated to a total of 13 sub-projects at the Agricultural University for the period 1971-1978.

The on-site disposal project started in 1972 and was managed by Hans Erik Stadshaug until 1974 and thereafter by Per Lindbak until 1978. Some studies were also conducted with septage lagooning (Anon, 1976). The research was then reorganized and funded by the Agricultural Research Council and the Norwegian State Pollution Control Authority (SFT) and coordinated as an inter-institutional project by the Section of Soil Pollution Research with Rolv Kristiansen as responsible coordinator.

Stadshaug started with an inspection of about 100 existing soil absorption systems and sand filter trenches (Stadshaug, 1974).

A. S. Eikum and R. W. Seabloom (eds.), Alternative Wastewater Treatment, 73–77.

Only a few of them were functioning satisfactorily because of extensive clogging mainly caused by lack of septage removal. For further studies 13 sand filters were chosen and the effluent from 5 of them was analyzed thoroughly. Very high purification efficiencies were found in these systems (Stadshaug, 1974). In the next phase, Stadshaug's group involved themselves in designing and consultant work, partly because of lack of knowledge throughout the country and partly because they could use the systems for research purposes. This included both sand filters and traditional absorption fields, some of them of considerable size. The field systems were taken over by Per Lindbak and the results were reported in an internal publication (Lindbak, 1977) and in a "PRA Users Manual" (Lindbak, 1978). Generally Lindbak reported unusually high purification efficiencies for sand filters. These results are, however, questionable because there was a certain dilution of the sand filter effluents caused by storm water and ground water.

Results reported by local authorities were of a much more pessimistic nature. It seemed as if there were a serious gap between the research results and practical experience with infiltration systems. This was due to lack of basic knowledge in the scientific field and lack of practical knowledge and management routines throughout Norway. It was felt that a new on-site research program was needed and the new program was started in 1980 as a three year program coordinated with an ongoing Swedish research program which Dr. von Brömssen will report on at this conference.

The aim of the joint research program is to increase knowledge about site selection, clogging of infiltrative surfaces and renovating processes in soil absorption systems. This information will be used to improve existing septic tank effluent manuals in Scandinavia. While the Norwegian research program mainly is concentrated on rural wastewater disposal, the Swedish program also comprises secondary effluent and storm-water.

PROJECTS IN THE NORWEGIAN RESEARCH PROGRAM

1. Site selection criteria for on-site disposal systems.
   Institution: Agricultural Research Council of Norway, Section Soil Pollution Research.

   Aim of Project: Development of usable methods for site selection and for evaluating the hydraulic capacity of soils for wastewater.

2. Microbial processes important to clogging and purification in soil absorption beds.
   Institution: Department of Microbiology, Agricultural University of Norway.
   Aim of project:
   a. To increase knowledge about microbial clogging of infiltrative surfaces.
   b. To optimize processes important to decomposition and turnover of pollutants in on-site soil absorption systems.

3. Removal of phosphorus in soil-filters for renovation of wastewaters.
   Institution: The Norwegian Forest Research Institute.
   Aim of project: To find a suitable phosphorus adsorption index for Scandinavian soil types. This will be used in soil selection procedures when choosing sites for on-site soil disposal of septic tank effluent.

4. Hygienic questions concerning removal of pathogens from percolating wastewater.
   Institution: Department Food Hygiene, Veterinary College of Norway.

Aim of project: The Norwegian part in this joint Swedish-Norwegian project is to investigate the travel-distances of microorganisms and parasites in different soil types loaded at varying rates, frequencies and temperatures.

5. Frost in septic systems.
Institution: Agricultural Research Council of Norway, Section Soil Pollution Research.
Aim of project: To evaluate needed insulation of soil absorption systems under different soil and climatic conditions.

6. Leachate treatment through soil percolation.
Institution: Agricultural Research Council of Norway, Section Soil Pollution Research.
Aim of project: To increase knowledge of necessary pre-treatment needs for leachate from small solid waste disposal sites before disposing in soil absorption fields.

7. Development of improved full-scale soil disposal systems.
Institution: Agricultural Research Council of Norway, Section Soil Pollution Research.
Aim of project: Based on in house studies and literature, improve existing types of on-site soil disposal systems in Scandinavia.

In addition, work on the improvement of biological toilets in progress since 1974 at the Agricultural University was to be correlated with the program. Today (1981) this work is concentrated on testing existing systems in the laboratory and in practice. A standard manual for testing biological toilets has been made. Each year the program makes a status report comprising results from the projects (Section Soil Pollution Research, 1981).

The staff working on the program includes geologists, microbiologists, a veterinarian, a soil chemist, a sanitary engineer and technicians.

DURATION OF PROGRAM

The first phase will be finished in 1982 when a "Scandinavian manual of septic tank practice" is expected to be published.

LITERATURE

1. Anon. 1978, "Jord som resipient". (Soil as recipient.) Report from research program for purification of sewage. NLH-Ås, 172 pp.

2. Lindbak, P.: 1977, "Avløp fra spredt bolig- og fritidsbebyggelse. Undersøkelser av rensesystemer basert på jord som rensemedium. Beskrivelse av de enkelte anlegg og deres funksjon". (Effluent from rural areas. Investigation of purification systems based on soil. Description of purification systems.) Ås-NLH. 296 pp.

3. Lindbak, P.: 1978, "Rensing av avløpsvann fra spredt bolig- og fritidsbebyggelse". (Purification of septic tank effluent from rural areas.) PRA 20. 54 pp.

4. Liseth, P.: 1980, "A Norwegian Research Program for Treatment of Waste Water (The PRA project). Paper presented at 8th CIB Congress, Oslo, June.

5. Stadshaug, H.E.: 1974, "Jord som rensemedium og resipient for kloakkvann fra enkelthus og mindre tettsteder." (Soil as a purification medium and recipient for sewage from single houses and rural settlings.) PRA-project 3.5. Infiltration of sewage and sludge. Ås-NLH. 37 pp.

6. Section Soil Pollution Research: 1981, Research program "Rensing av avløpsvann i jord - spredt bosetning". (Research program Purification of septic tank effluent in soil - rural areas.) Annual report 1980 with planned activities for 1981 and 1982.) Report No. 2. Ås-NLH. 88 pp.

RESEARCH ON ON-SITE DISPOSAL METHODS IN SWEDEN - PAST AND PRESENT

Ulf von Brömssen

Fil. lic., The National Swedish Environment Protection Board, the Research Department (SNV - Ff), Box 1302
171 25 Solna, Sweden

SUMMARY

In 1977 SNV - the National Swedish Environment Protection Board decided to initiate a 5-year project with a budget of approximately US$ 250,000 each year for research on soil processes, transport and reduction in soil of pollutants in connection with controlled and uncontrolled infiltration of polluted water.

The ultimate aim of this research is:

- to get sufficient information to base new guidelines for the protection of the ground water resources in Sweden.
- to get a better understanding of the processes in the soil in connection with infiltration, in order to issue new manuals for subsurface on-site systems.
- to get better information on the effects on ground water quality caused by infiltration of urban storm water into subsurface infiltration systems.
- to get better information on leachate from municipal household solid waste deposits, including:
  possibilities to reduce the formation of leachate,
  possibilities to control the chemical composition of leachate,
  possibilities of pretreatment of leachate before infiltration.

A. S. Eikum and R. W. Seabloom (eds.), Alternative Wastewater Treatment, 79–90.

The paper gives a background of Swedish guidelines and research on on-site infiltration systems and a presentation of the research in connection with the Swedish infiltration project. Most of the papers produced so far are written in Swedish. In order to widen the information on this project and facilitate contacts, postal addresses are given in connection with the projects.

The paper also urges a systematic publication program to spread the scientific results and to apply the results to specific practical problems. This information should be addressed not only to other scientists, but to decisionmakers, creators of public opinion, consultants and contractors.

INTRODUCTION

In Sweden with its population of 8 million people, the use of on-site wastewater systems for single households increased in 1962, when the first guidelines on this subject were issued. During the following 15 years attempts were made to develop the design of subsurface infiltration trenches and drained sand filter beds. Efforts were also made to determine the reduction rate of BOD, SS, phosphorus, nitrogen and pathogens. Most of these investigations had the black box approach, i.e. analyzing water in and water out. In 1977 the National Swedish Environment Protection Board (SNV) decided to initiate a 5-year project for research on soil processes, transport and retention in soil of pollutants in connection with infiltration of polluted water in soil. The aim being to get information on soil processes, design criteria for subsoil systems and technical background information on issuing new regulations for the protection of the ground water resources in Sweden.

This paper deals with the background of this project and the presentation of the projects undertaken by the Swedish infiltration project.

In Figure 1 is given the situation for water supply and sewage treatment in Sweden in 1979. Notice that almost 50 percent of the population is dependent on groundwater and that 3.7 million people, permanent living and in recreation houses, use groundwater and usually also sub-surface infiltration systems for their sewage.

Type of urbanisation	Water supply Million people		Sewage treatment Million people		
	Ground water	Surface water	Biological and/or chemical treatment	Septic tank only	Non specified x)
Municipalities	3.2	3.6	6.6	0.2	-
Rural areas	1.4	-	-	-	1.4
Recreation houses (650,000)	2.3	-	-	-	2.3
TOTAL	6.9	3.6	6.6	0.2	3.7

x) Usually septic tanks mostly in combination with sub-soil infiltration systems.

Figure 1. Water supply and sewage treatment in Sweden 1979. Notice that 8.2 million represent permanent living and that out of these 2.3 million also utilize recreation houses.

EARLIER AND PRESENT REGULATIONS

With the advent of urbanization and implementation of municipal sewage treatment, the proper disposal of wastewater from rural homes has been advocated. In 1950 national regulations were issued on septic tanks as being the best alternative for sewage treatment for single households. These regulations were issued in connection with a government survey SOU 1955:18 (Statens Offentliga Utredningar).

In 1962 the first guidelines were issued on small scale wastewater infiltration systems for single houses. (Med. från KVVS, No. Va 8 (Kungliga Väg och Vattenbyggnadsstyrelsen).) These guidelines were later reprinted in 1974 by the National Swedish Environment Protection Board (SNV) which was set up in 1967. The guidelines are found in the new SNV series as publication: SNV Publ. 1974:15.

The 1962/74 guidelines did not deal with the rate of reduction of pollutants, nor did it deal with design criteria regarding hydraulic and pollutant load. All the designs so far have been based on the assumption that the load of pollutants is that of household water.

The guidelines state that the septic tank for combined gray and black water should be 2,000 liter (wet volume) when emptied once a year.

The infiltration trench should be 3.5 m/person or 7.0 m/p according to the sieve analysis of the soil. On-site systems should be designed for minimum 4 persons. Distribution pipes should be minimum 100 cm above the highest groundwater level and the minimum depth of soil below the same pipes should be 100 cm. In case of finegrained soils or high groundwater levels a drained sandfilter bed should be used.

PAST AND PRESENT RESEARCH

The 1962 guidelines initiated research on the characteristics of sewage water as well as laboratory and full scale investigations on on-site systems, in order to find out reduction rates of different parameters. The greatest interest has been directed toward the reduction of phosphorus, in order to influence the authorities to allow a greater use of sand filter beds. The authorities are very concerned about limiting phosphorus in the surface receiving waters.

Some of the research projects between 1962 and 1977 are listed below:

- Determination of the characteristics of household sewage water (Ahl, T. et al. 1963).
- Pollution of ground water from on-site infiltration systems (Carlsson & Horkeby 1976, Carlsson, L. 1977).
- Studies of reduction and retention of on-site installations (Nilsson, K. 1973, Nilsson, P. 1976).
- Laboratory and site studies on on-site infiltration systems (Nilsson, K. 1977).

This research did not investigate fundamental processes, but was rather more of the black box concept. Water in and water out from on-site systems was analyzed, parameters in the system were altered and the effects were recorded. In 1977 SNV initiated the earlier mentioned 5-year project intitled "Infiltration-processes in the soil and ground water pollution control". This project deals with fundamental soil processes, transport and retention of pollutants in the soil.

The wastewaters studied are not limited to household sewage water, but do also include studies on leachate from sanitary land-fills and urban storm water.

All this research deals with the fundamental processes in the soil as well as attempting to answer different questions connected with the design and management of sub-soil systems, uncontrolled infiltration and ground water pollution control. Close collaboration on this research has taken place between the Nordic countries, especially between Norway and Sweden. Several of the project members have also cooperated with US colleagues.

The projects listed below have been undertaken by different institutions, but financed by SNV within the Infiltration Project.

To facilitate contacts, some postal addresses are given in connection with the names of the investigators.

LIST OF PROJECTS

- Microbiological processes in sand filter trenches. Investigations to present a pilot scale denitrification sand filter system.

  In this project an engineering institution on sewage processes works in collaboration with a microbiological institution. A pilot scale sand filter system has been built and loaded with inoculated artificial sewage water. Microbiological processes are studied by counting relevant bacteria and at the same time analyzing the changes of BOD, tot-P, tot-N, NH_4, NO_3, NO_2, and N_2 in the sand filter.

  Special interest is focused on the creation of denitrification in sand filter trenches. The aim is to design a simple system which works with denitrification by adding raw sewage water to treated nitrified sewage under anaerobic conditions.

  Fred Nyberg, civil engineer at the Royal Institute of Technology in Stockholm.
  Address: The Royal Institute of Technology, Water Resources Engineering, 100 44 Stockholm, Sweden.

  Michael Andersson, microbiologist at the Agricultural University.
  Address: Department of Microbiology, The Swedish University of Agricultural Sciences, 750 07 Uppsala, Sweden.

- Adsorption of heavy metals to Swedish soils. Metals studied are Zn, Cu, Pb, and Cd. The Langmuirisoterm has been used to describe the adsorption capacity of sandy soils, till and clay.

  Arne Andersson,
  Address: Department of Soil Sciences, The Swedish University of Agricultural Sciences, 750 07 Uppsala, Sweden.

- Studies on pond infiltration as a final step for municipal treatment plants. Infiltration of mechanically, biologically and chemically pretreated sewage has been studied. A research station has been built at a treatment plant, where as the last step infiltration into open ponds is used. Sewage water treated at different stages has been used in column studies. (Ø 300 mm, length 2,000 mm). The hydraulic loads are (exclusive resting periods) for the full scale infiltration 8 - 10 cm/h, for column studies 2 - 12 cm/h.

Peter Nilsson, civil engineer
Address: Department of Water Resources Engineering,
Lund Institute of Technology, University of Lund,
Box 725, 220 07 Lund, Sweden.

- Development of additives to increase the adsorption of phosphorus in sub-surface on-site systems. Industrial rest-products are being studied. The aim is to develop a filter unit, which can be exchanged when no more phosphorus is adsorbed.

Peter Nilsson, civil engineer. Address: See above.

- Survival and transportation of microbial pathogens and indicators in soil and ground water.
  The aims of the project are to give a better understanding of the different biological, chemical and physical factors that interact with allochtonous sewage microorganisms, and how the survival and percolation of these organisms are influenced.

  This work will be carried out in the laboratory, as well as in half scale and full scale plants. Parameters like types of soil (dominant in Sweden), temperature, inorganic ions, organic material, antagonism and predation are of prime interest, and these as well as clinical isolate will be tested.

They will also be characterized with regard to surface properties as hydrophobicity and surface charge. The hygienic risk of waterborne diseases is normally based on microbial indicators, like the coliform organism, or more specificly E. Coli. In order to get a more differentiated picture, other presumptive indicators like fecal streptococci, anaerobic bacteria (Clostridium sp., Bifidobacterium and Bacteroides) and bacteriophages are to be studied (these have not up to now been used in Sweden). Specific chemical indicators, like fecal sterols, are also studied.

The project is coordinated with a similar project in Norway, run by the University of Agriculture in Ås and Department of Food Hygiene, Veterinary College, Oslo, and in parts also with Danish activities.

Tor Axel Stenström and Sven Hoffner, microbiologists
Address: The National Bacteriological Laboratory,
Department of Water Microbiology, Fack 105 21 Stockholm,
Sweden.

- Calculation of transport of pollutants in ground water - processes - quantifications - modelling.

The aim is

1. Presentation of a state of the art report on transport, adsorption and distribution of pollutants in ground water with special emphasis on diffusion-dispersion.
2. Modelling of ground water flow in typical Swedish geological environments (velocities, flow patterns and influence from discharge and recharge).
3. Discussion of input factors in modelling calculation of transport of ground water contaminants.

Leif Carlsson and Anders Carlstedt
Address: Geological Survey of Sweden,
Box 670, 751 18 Uppsala, Sweden.

- Modelling of transport of contaminants in ground water.

  A mathematical finite element model has been modified and made available on computer. The model takes into account transport parameters such as convection, dispersion and adsorption. The model has been run on three actual field cases and the results are discussed.

  Clifford Voss and Hans Hydén
  Address: The Royal Institute of Technology, Department of Water Resources Engineering, 100 44 Stockholm, Sweden.

- Field studies of ground water contamination in connection with infiltration in subsoil systems with high ground water levels.

  The required distance of a minimum of one meter from the distribution pipe to the highest ground water level can not be obtained in all parts of Sweden, due to climatic and geological conditions. It is therefore of interest to know how this will affect the ground water quality. A great number of single house infiltration systems have been investigated to find installations where the ground water is high and where, due to this, the sewage is flooded in the distribution pipes.

  Fred Nyberg
  Address: See above.

- Studies to improve the functioning of large 3 chambered septic tanks. Uncontrolled transport of floating sludge from the septic tank creates problems to the sub-surface systems. It is of great importance to minimize this transport. A newly developed, standardized test technique for septic tanks will be used.

  Fred Nyberg
  Address: See above.

- Background relationships and suggested critical limits of phosphorus in lakes and streams.

  Which types of receiving waters are suited or not suited for effluents of treated sewage from sand filter beds? What are the biological effects on the water environment at different concentrations of phosphorus in the water? How should the natural background concentration be calculated? What are the criteria for critical limits of phosphorus? These questions are dealt with in a paper presented in Swedish.

Torgny Wiederholm
Address: The Swedish Environment Protection Board, Water Quality Laboratory Uppsala, Box 8043, 750 08 Uppsala, Sweden.

Eugene B. Welch
Address: University of Washington, Department of Civil Engineering, Environmental Engineering & Sciences, FX-10, Seattle, Washington 98195, USA.

- Bacteriological investigations of drainage water from church yards. Accumulation of Pb and Cu in soil from run off from church roofs.

  Ground water contamination from church yards is a subject of great concern for the public. This is the case when ground water wells are situated downstream from a church yard or when drainage water is discharged into a lake or a stream. In this project bacteriological samples have been taken from ground water and from drainage water from church yards. Tracer studies have been performed to see if added bacteriophages and antibiotic resistant bacteria could be traced in the collected water samples.

  The second part of the project deals with soil sampling and analysis of Pb and Cu at places, where water from church roofs has been infiltrated for long periods of time. This is an attempt to investigate long term effects of heavy metal disposal in soils and ground water from infiltration of storm water. This information will compliment another project which deals with infiltration of storm water in three types of areas: a residential area, an industrial area, and a highway.

Jan Rennerfelt, professor
Address: Scandiaconsult, Box 4560, 102 65 Stockholm, Sweden.

In addition to these projects studies have been performed on ion exchange in bentonite, when in contact with leachate from solid waste. Case studies have been carried out on ground water contamination from storm water infiltration in urban, industrial and motorway environments.

When outlining the research programs in Norway and Sweden, it was agreed that in order to better utilize the limited resources, some projects should be carried out in Norway and some in Sweden. Part of the Swedish research program is therefore carried out in Norway only. These projects are:

- Frost penetration in winter time - effects on sub-soil systems
- Procedures for the determination of infiltration capacities (clean water) in Scandinavian soils
- Hydraulic studies on the distribution system for sub-soil systems
- Determination of a phosphorus adsorption index for soils
- Basic studies on phosphorus retention in soils.

FINAL REMARKS

The Swedish infiltration project will be terminated in June 1982. It is of great importance that the scientific results be presented not only to other scientists, but also to those who in their daily life work with practical aspects of sewage treatment. This means for example, that certain research results will have to be applied to and discussed in connection with well defined practical problems. This calls for a prolonged dialogue between the scientists and the users of the scientific results, a dialogue which must be initiated and followed up by the coordinators of the research projects. It is therefore important that funds are made available for the proper dissemination of this information.

SESSION IV

THE SOIL AS A RENOVATING MEDIUM

Chairmen: R.W. Seabloom
J.F. Kreissl
A.S. Eikum

WATER MOVEMENT INTO AND THROUGH SOIL

Richard J. Otis, P.E.

Professor of Civil & Environmental Engineering
University of Wisconsin, Madison, Wisconsin 53706,
USA

INTRODUCTION

All homes in unsewered areas must have a safe and effective means of wastewater disposal. Where soils are suitable, subsurface soil absorption of septic tank effluent is the most reliable and least costly. This is due to the soil's very large capacity to transform and recycle most pollutants found in domestic wastewaters.

Soil suitability is dependent upon its capability to absorb and purify the wastewater. System failure occurs if either of these functions is not performed. Both are related directly to the hydraulic conductivity of the soil which is controlled largely by the pore geometry of the material. Therefore, an understanding of how water moves through the soil is necessary to predict the potential of the soil for wastewater absorption and treatment.

PHYSICAL PROPERTIES OF SOIL

Soil is a complex arrangement of solid particles and voids or pore spaces. Water movement through soil must occur within the pore spaces. Therefore, the size, shape and continuity of

A. S. Eikum and R. W. Seabloom (eds.), Alternative Wastewater Treatment, 93–104.

the pore spaces are very important. These characteristics are dependent primarily on the texture, structure, bulk density and clay mineralogy of the soil.

Soil Texture

Texture refers to the relative proportion of the various sizes of solid particles smaller than 2 mm in diameter that make up the soil. The particles are classified into three size fractions called soil "separates". They are sand, silt and clay. The size limits for each separate are presented in Table 1. Twelve textural classes used to describe soils are defined by the relative proportions of the separates by weight. These are represented in the "textural triangle" shown in Figure 1.

Aeration and drainage of the soil are related closely to its texture because of the influence particle size has on the pore size and pore continuity. Clay soils are not very permeable because the small particles result in very fine, discontinuous pores. Sands are much more permeable (though less porous) because the pores between the grains are large and continuous.

Soil Structure

Structure refers to the relative arrangement of the solid particles to one another. It also affects the size and shape of the soil pores. In granular soils such as sands, the pores are simply packing pores between the individual grains. The size and shape of these pores is primarily a function of texture, shape and packing of the individual grains. In soils with significant amounts of clay and organic matter, the soil particles become cemented together to form aggregates or "peds". Surfaces of weakness separate the peds and are often seen as cracks in the soil.

Table 1. Size Limits for Soil Texture Separates (U.S. EPA, 1980).

Soil Separate	Size Range mm	Tyler Standard Sieve No.
Sand	2.0 - 0.05	10 - 270 mesh
(very coarse)	2.0 - 1.0	10 - 16 mesh
(coarse)	1.0 - 0.5	16 - 35 mesh
(medium)	0.5 - 0.25	35 - 60 mesh
(fine)	0.25 - 0.10	60 - 140 mesh
(very fine)	0.10 - 0.05	140 - 270 mesh
Silt	0.25 - 0.002	-
Clay	< 0.002	-

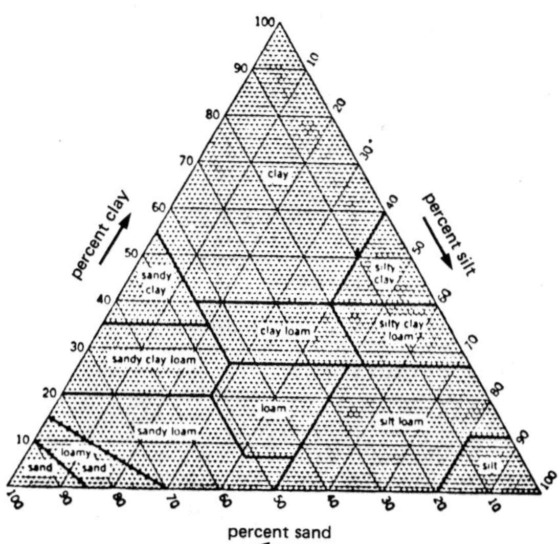

Figure 1. Textural Triangle Defining the Twelve Textural Classes. (Soil Conservation Service, 1951.)

The type of structure is defined by the shape and arrangement of the peds. The seven basic types are described in Table 2 and illustrated in Figure 2. These structure units may be altered or destroyed due to changes in moisture content, chemical composition of the soil solution, biological activity and management practices. Soils containing minerals that shrink and swell appreciably, such as montmorillonite, show particularly dramatic changes.

The effects of structure on the soil's permeability are significant. The planar voids between the peds often are relatively large and continuous compared to the voids between the primary particles within the peds, increasing the soil's permeability over that which would be expected based on soil texture alone. The type of structure also determines the dominant direction of the pores and, hence, the direction of water movement. Platy structures restrict vertical percolation, prismatic and columnar structures enhance vertical percolation, and blocky and granular structures enhance percolation both horizontally and vertically.

Bulk Density

Soil bulk density is the ratio of the mass of the soil to its bulk or volume. Of soils with the same texture, those soils with higher bulk densities are more dense with less pore volume and, therefore, less permeable.

Clay Mineralogy

Some clay minerals such as montmorillonite shrink and swell appreciably with changes in soil moisture. Even small amounts of swelling clays can affect the soil permeability dramatically due to large cracks that open and close with wetting cycles.

Table 2. Description of Soil Structure Types.

<table>
<tr><th colspan="2">STRUCTURE TYPE</th><th colspan="2">DESCRIPTION</th><th>HORIZON</th></tr>
<tr><td colspan="2">Granular</td><td>Non-porous peds</td><td rowspan="2">Nearly spherical with many irregular surfaces</td><td rowspan="2">Usually found in surface soil or A horizon</td></tr>
<tr><td colspan="2">Crumb</td><td>Porous peds</td></tr>
<tr><td colspan="2">Platy</td><td colspan="2">Plate-like aggregates with horizontal axis greater than vertical. Plates usually overlap.</td><td>Usually found in subsurface or A_2 horizon</td></tr>
<tr><td colspan="2">Angular blocky</td><td>Ped faces flattened and sharply angular vertices</td><td rowspan="2">Block-like with three dimensions nearly equal around a point</td><td rowspan="4">Usually found in subsoil or B horizon</td></tr>
<tr><td colspan="2">Subangular blocky</td><td>Mixed rounded and flattened faces with many rounded faces</td></tr>
<tr><td colspan="2">Prismatic</td><td>Without rounded caps</td><td rowspan="2">Prism-like with the vertical axis greater than the horizontal</td></tr>
<tr><td colspan="2">Columnar</td><td>With rounded caps</td></tr>
<tr><td rowspan="2">Structure lacking</td><td>Single grain</td><td colspan="2">Soil particles exist as individuals not forming aggregates</td><td rowspan="2">Usually found in parent material or C horizon</td></tr>
<tr><td></td><td colspan="2">Soil material clings together in large uniform masses</td></tr>
</table>

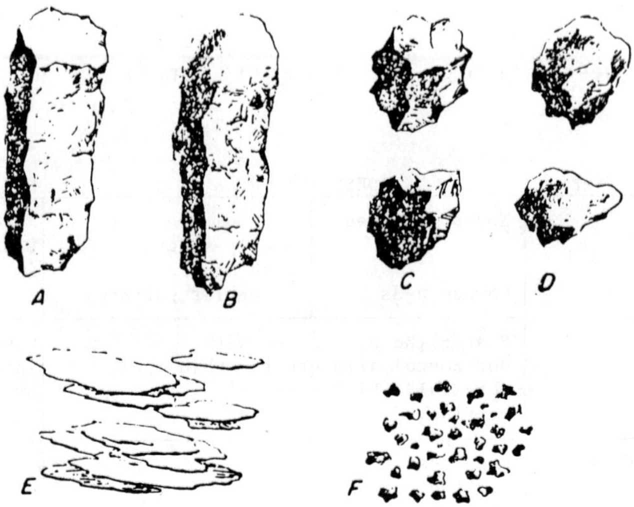

Figure 2. Illustration of Basic Soil Structure.
Types: (A) Columnar, (B) Prismatic,
(C) Angular Blocky, (D) Subangular Blocky,
(E) Platy, (F) Granular or Crumb.

WATER MOVEMENT IN SOIL

The flow of water results when there exists a difference between the potential energy status of water within the soil. The direction of flow is toward decreasing potential. The rate of flow is proportional to the potential gradient and is affected by the geometric properties of the pore channels through which the flow takes place.

Direction of Water Movement

Water moves from a point of higher potential energy to a point of lower potential energy. The energy status of water in soil is referred to as the "soil moisture potential". It has several component potentials of which the "gravitational" and "matric" potentials are the most significant in wastewater disposal.

Gravitational potential. The gravitational potential is due to the earth's gravity. It is determined at any point by the elevation of that point relative to a fixed reference elevation.

Matric potential. The matric potential is produced by the affinity of water molecules to each other and to solid surfaces. This affinity results in the phenomenon of capillary rise. The soil pores act as capillary tubes, drawing water into them (see Figure 3). Since the water is held against the force of gravity, it has a pressure less than atmospheric. This negative pressure is often referred to as "soil suction" or "soil moisture tension". Increasing suction or tension is associated with soil drying. The matric potential is zero when the soil is saturated.

The moisture content of soils with similar moisture tensions varies with the nature of the pores. Water is held tighter in smaller pores than in larger pores. Upon draining, the largest pores will empty first because they have the weakest hold on the water. Therefore, in unsaturated soils, the water is held in the finer pores. Figure 4 illustrates the change in moisture content versus changes in moisture tensions. Because sand has many relatively large pores, it drains abruptly at relatively low tensions, whereas clay releases only a small volume of water over a wide tension range because most of it is retained in the very fine pores which are characteristic of clay. The moisture retention curves of the other two soils lie between the sand and clay curves because the soils have more fine pores than the sand, but fewer than clay.

Rate of Water Movement

Water movement in soil is dependent on the soil's hydraulic conductivity, or its ability to transmit water and a moisture

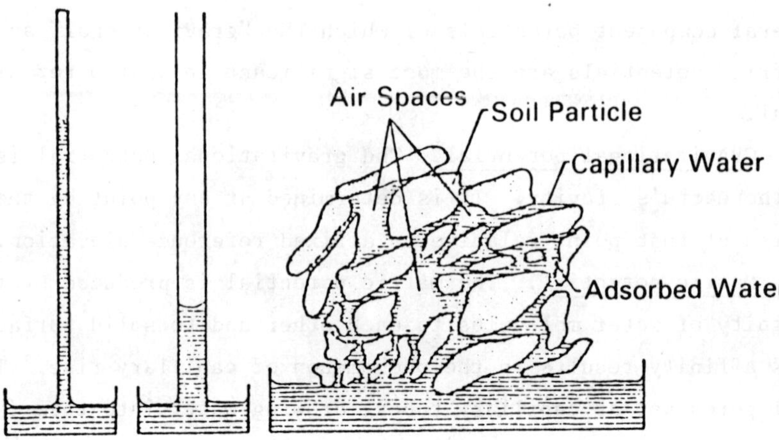

Figure 3. Upward Movement by Capillarity in Glass Tubes as Compared with Soils (After Brady, 1974).

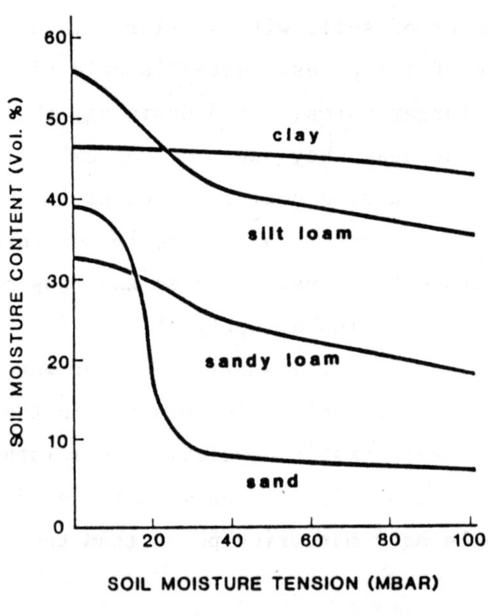

Figure 4. Moisture Retention Curves for Four Different Soil Textures (Bouma et al., 1972).

potential gradient to provide a driving force. The hydraulic conductivity of the soil is related to the number, size and geometry of the soil pores. Soils with large continuous water-filled pores can transmit water more easily and therefore have a higher hydraulic conductivity than soils with small, discontinuous water-filled pores. It can change dramatically with changes in the soil moisture tension. At a tension equal to or less than zero, the soil is saturated and all the pores are conducting the water. When the tension is greater than zero, air is present in some of the pores and unsaturated conditions prevail. This condition grossly alters the flow channel because the forces which cause the flow, are now associated with capillarity. As the water content decreases or tension increases, the path of the flow of water becomes more and more tortuous. Therefore, the unsaturated conductivity is usually much lower than the saturated conductivity (see Figure 5).

The greater the difference in the total potential between two points, the more rapid the movement. However, the volume of water moved over a given time period is proportional to the total potential gradient and the soil's hydraulic conductivity at the prevailing moisture content as defined by Darcy's Law:

$$Q = KA \frac{dh}{dz}$$

Where: Q = flow rate

K = hydraulic conductivity

A = cross-sectional area of flow

$\frac{dh}{dz}$ = total potential gradient.

The hydraulic conductivity, K, decreases with increasing soil moisture tension (Figure 5). The rate K decreases with increasing moisture tension is characteristic for a given soil. Coarse soils with predominantly large pores (such as sands) have relatively high hydraulic conductivities at saturation, but drop

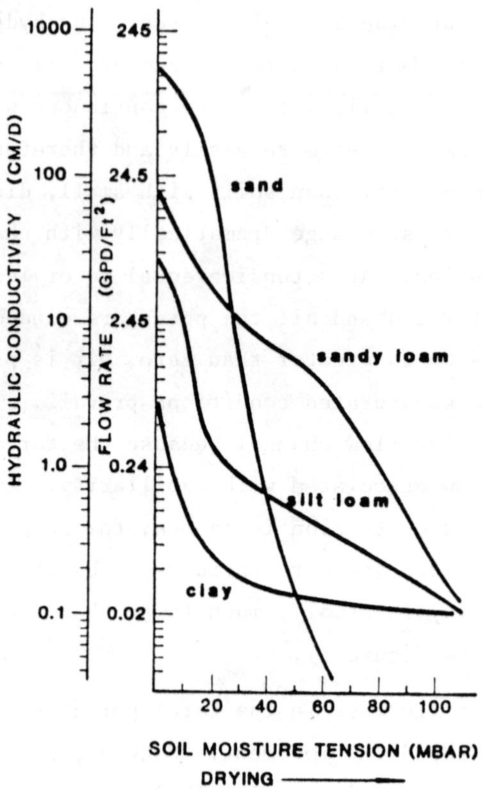

Figure 5. Hydraulic Conductivity Versus Moisture Tension. (Bouma, 1975.)

off rapidly with increasing moisture tension as the large pores empty. Fine textured soils with predominantly small pores have relatively low conductivities at saturation, but their conductivities decrease more slowly upon increasing tension. The effects of structure on the conductivities are illustrated by the clay curve in Figure 5. The planar voids between the peds contribute much to the flow rate near saturation, but they are the first to drain with increasing tension. As a result, the hydraulic conductivity drops rapidly at first and then levels off as the finer pores begin to control the flow.

Flow Through Layered Soils

Soil layers of varying hydraulic conductivities interfere with vertical water movement. Abrupt changes in conductivity can cause the soil to saturate or nearly saturate above the boundary regardless of the hydraulic conductivity of the underlying layer. If the upper layer has a significantly greater conductivity, the water ponds because the lower layer cannot transmit the water as fast as the upper layer delivers it. If the upper layer has a lower conductivity, the underlying layer cannot absorb it because the finer pores in the upper layer hold the water until the matric potential is reduced to near saturation.

SIGNIFICANCE OF UNSATURATED FLOW

With continuous wastewater application to the soil, it is common to develop a clogging mat with a hydraulic conductivity lower than that of the underlying soil. The clogging mat may restrict water movement to a degree where water is ponded above while the soil below remains unsaturated. Water passes through the mat due to the potential gradient created by the hydrostatic pressure of the ponded water pushing the water through, and the capillarity of the soil below wicking the water through.

Though the unsaturated condition created in the soil by the clogging mat slows the rate of infiltration, it enhances treatment. Wastewater is purified by filtration, biochemical reactions and adsorption, processes which are more effective in unsaturated soils because the average distances between wastewater pollutants and the soil particles are decreased and the contact time increases.

REFERENCES

Bouma, J., W.A. Ziebell, W.G. Walker, P.G. Olcott, E. McCoy & F.D. Hole: 1972, "Soil absorption of septic tank effluent, a field study of some major soils in Wisconsin." Information Circular No. 20, Geological and Natural History Survey, University Extension, Madison, Wisconsin.

Bouma, J.: 1975, "Unsaturated flow during soil treatment." Journal of the Environmental Engineering Division, ASCE, 101: pp. 967-983.

Brady, N.C.: 1974, The Nature and Properties of Soils. 8th edition. MacMillan, New York. 655 p.

Soil Conservation Service: 1951, Soil survey manual, USDA Handbook 18. U.S. Government Printing Office, Washington, D.C.

U.S. Environmental Protection Agency: 1980, "Onsite wastewater treatment and disposal systems: design manual." Technology Transfer, Cincinnati, Ohio.

THE SOIL AS A RENOVATING MEDIUM - CLOGGING OF INFILTRATIVE SURFACES

Rolv Kristiansen

Research Coordinator, Agricultural Research Council
of Norway, Section Soil Pollution Research,
1432 Ås - NLH, Norway

INTRODUCTION

One of the most investigated processes of on-site soil absorption systems is the clogging phenomenon. In spite of this, clogging or biocrust formation is still a mysterious process regarded by most sanitary engineers as a serious disadvantage to septic systems. Scientists working with on-site absorption systems have, however, different opinions about this very important part of every absorption field.

A common misunderstanding about clogging is the assumption that the soil surface gets completely blocked. This is not true. It is rather a reduction of the soil pore-volumes and thereby a reduction of the hydraulic capacity of the field. Water is still flowing into the soil, but at a considerably lower rate than is possible in virgin soil.

Basically, there are several reasons for clogging. First, shearing and compaction of soil surfaces by heavy machinery may cause clogging (McGauhey et al., 1966). A second type which is common in lagoons, is clogging caused by algae. This paper will discuss the type of clogging zone which develops in absorption

A. S. Eikum and R. W. Seabloom (eds.), Alternative Wastewater Treatment, 105–120.

beds laid underground in permeable soils not damaged by machinery etc. In septic systems clogging usually occurs after a few months' loading and is the result of physical, chemical and biological factors.

THE CLOGGING PHASES

In the literature, the clogging process is often divided into two or three phases (Allison, 1947; Jones & Taylor, 1965; Kristiansen, 1981 a; Magdoff & Bouma, 1974; Thomas et al., 1966). When three phases are described, one phase can be explained by entrapped air which has to be removed before water can flow freely through the soil medium. This phase will probably last only a very short time and is therefore of minor importance to the ultimate hydraulic capacity of the system. A more simple and probably more realistic way of illustrating the clogging process is without the entrapped air phase (Figure 1). Thus the first phase is characterized by a rapid decrease in infiltrative capacity followed by a very slow reduction in the acceptance rate which will probably reach some sort of equilibrium. Laak (1980) mentions it as the "LTAR" value or "long term acceptance rate". There is, however, a certain amount of controversy regarding the stability of this long term acceptance rate, but the half-life for systems installed in most soil types is found to be more than 30 years (Hill & Frink, 1980).

THE CAUSES OF CLOGGING

What does the clogging zone contain? Very few investigations have been conducted on the composition of the clogging zone. The zone is mostly characterized as a "black slimy layer" in the infiltrative surface. The black zone can, however, be seen at greater depths than the slime layer where most of the hydraulic resistance is found. One has to separate the deep penetrating black zone from the real crust zone. Investigations have shown

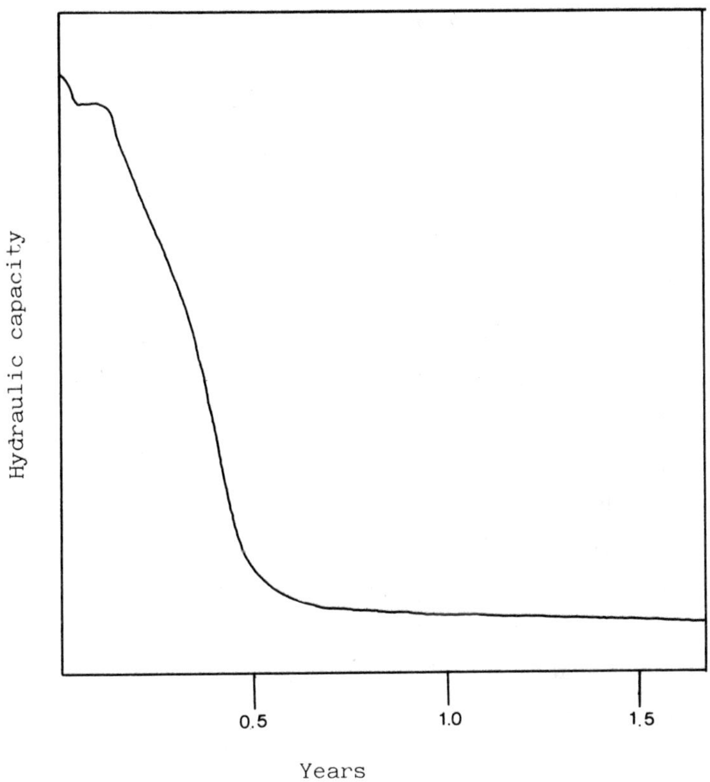

Figure 1. Typical decrease of hydraulic capacity of an infiltrative surface.

that the gravel/soil interface contains large amounts of organic material while the lower black zone contains FeS which is of minor importance to the hydraulic capacity. FeS is only an indicator of anaerobic conditions (Kristiansen, 1981 a; Laak, 1981; Thomas et al., 1966).

The organic C analyses of soil from septic systems show a considerable accumulation of organic material in a thin zone between gravel and underlying soil (Kristiansen, 1981 a; Walker et al., 1973). This is in accordance with the fact that the water restricting layer only penetrates 0.5-5 cm into the soil as proved by tensiometer readings (Bouma et al., 1972). It seems reasonable to assume that the clogging mat mainly is composed of accumulated suspended solids (SS), bacterial cells and fragments of microorganisms. These are mostly biodegradable. The environment in the clogging zone is, however, highly reduced (Figure 2). Under such conditions there will only occur a partial decomposition of organic material and a certain humification will occur.

The amount of polysaccharides and polyuronides has been shown to correlate with the degree of clogging (Avnimelech & Nevo, 1964; Mitchell & Nevo, 1964). It is probable that suspended matter may be degraded into polyuronides under anaerobic conditions and thereby aggregate soil particles. Mitchell & Nevo (1964) also pointed out that bacterial cells accounted for 10 percent of the clogging effect. This was assumed on the basis of plate counts which later have been shown only to give about one percent of the actual number of bacteria in septic systems (Kristiansen, 1981 c). Microscopic counts of sand from a sand filter showed that the maximum pore volume filled with cells was only 1.4 percent. It is probable that the real biovolume is much greater since most soil bacteria produce extracellular slimy material. In nature bacteria are covered by a "glycocalyx" of fibers that adhere to surfaces and to other cells (Costerton et al., 1978). These fibers may be detected by specific staining and electron-microscopical investigations, and in the opinion of the author, these structures play a significant role in the ultimate phase of the clogging process. Further research needs to be done from a very basic point of view before the answer is available.

In summary, the first phase of the clogging process is caused by the accumulation of suspended solids (SS) and the next or ultimate is caused by a bridging of the SS particles/soil particles by bio-produced material which slowly accumulates and after some time gets into some sort of steady state.

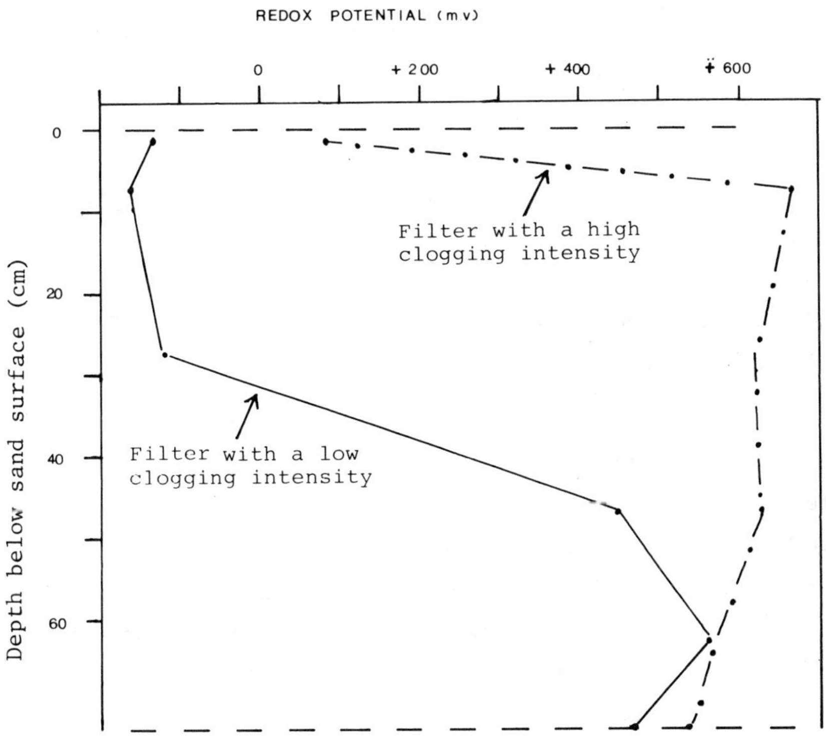

Figure 2. Redox potential in a sand filter with high and low clogging intensity.

The break-throughs which may be observed in column experiments are caused by a breaking up of the "bio-bridges" due to the death of micro-organisms. In a soil absorption field there is probably a continuously "build-up-breakthrough process" going on, which plays an important role for the long term acceptance rate.

EFFECT OF LOADING CONDITIONS ON CLOGGING

Continuous trickling of septic tank effluent into soil absorption systems is often said to enhance clogging because of the local overloading of the soil. It is the most usual and simple loading type which gives this so-called creeping clogging

(Bouma et al., 1972). Equal distribution of the wastewater by a dosing system will prevent creeping clogging, but the question is; will it prevent clogging? Comparable field experiments have not been done on this, but columns with an equal distribution do clog. The writer feels that clogging will occur in both cases, but it may take longer time before ponding of the infiltrative surface occurs when water is distributed over the whole surface. In addition, the question of an equal distribution or not is more a question of renovation capacity than hydraulic capacity. This will be discussed.

In one study with sand filter columns loaded with septic tank effluent at a rate of 5 cm/day different dosing intervals were compared (Stavn & Kristiansen, 1981). Dosing twice a day gave a much higher reduction in hydraulic capacity than dosing every 20 minutes, which is close to a continuous loading. These effects could only be seen the first six to seven months of the experiments. After 10 months no effects of loading frequencies could be detected.

Increasing the total hydraulic loading, however, decreases time to ponding of the infiltrative surface. Figure 3 shows reduction in hydraulic capacity as a function of the total loading of graywater on sand filter columns.

EFFECT OF TYPE OF WASTEWATER ON CLOGGING

In Allison's (1947) classical experiments, clogging occurred in columns loaded with tap-water, but not with tap-water to which mercuric chloride was added. Similar results were reported by Gupta & Swartzendruber (1962) when phenol was added. These effects are generally explained by microbial activity and lead to the conclusion that a certain degree of clogging must always be expected when nonsterile water is infiltrated in soil.

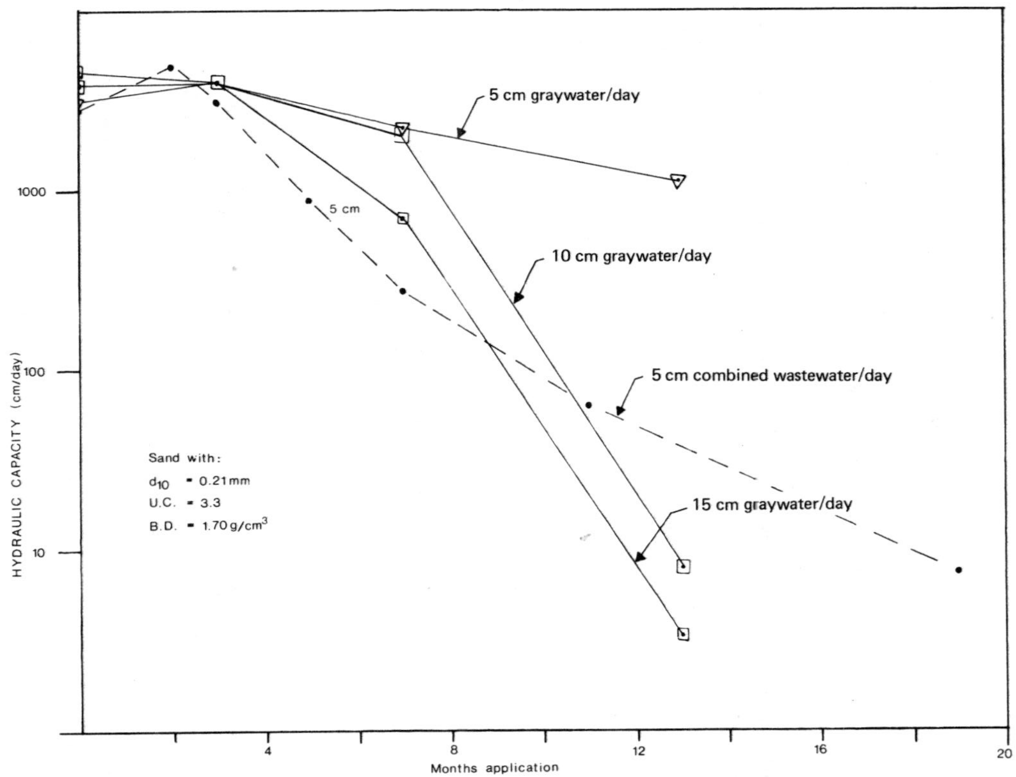

Figure 3. Hydraulic capacities of sand filter columns loaded at different rates of graywater and 5 cm combined septic tank effluent. Each point is average of three columns.

It seems reasonable to assume that a higher degree of pretreatment will reduce clogging and thereby reduce the area required for proper absorption. Laak et al. (1974) proposed that the infiltrative area could be adjusted according to the following formula:

$$\text{Adjusted area required} = \text{Area required for standard septic tank pretreatment} \sqrt[3]{\frac{BOD_5 + TSS}{250}}$$

The National Environmental Health Association (1979) mentions some cases where aerobic effluent is more readily and continuously accepted than septic tank effluent, while EPA (1978) states that the effect of effluent quality on soil clogging is uncertain. These contradictions may be explained by investigations with different wastewater and soil types.

Aerobic unit failure may have dramatic effects on the absorption field because of sludge. Suspended solids in effluent from an extended aeration unit clogged a fine soil more heavily than septic tank effluent (Daniel & Bouma, 1974). This was explained by the different sizes and shapes of SS in the two types of effluent. It was suggested that finely divided particles in the aerobically treated effluent penetrate the relatively porous topsoil more easily and form what the authors called "bottlenecks" in the pore systems.

EPA (1978) suggests that "some substance in wastewater which is not present in tap-water causes soil clogging and that the substance can be partly reduced by aerobic pre-treatment". The effects of aerobic pre-treatment are, however, only seen in coarse soils. The author doubts this, and feels that the explanation is more simple. It is possible that a higher content of organic material in the wastewater will give heavier biological growth in the crust zone in addition to the physical entrapment of SS in the soil.

In recent years column studies performed in Sweden with different types of secondary effluents have shown that the only type of pre-treatment that may allow a reduced absorption area compared to a standard septic system is precipitation (P. Nilsson, 1981).

In practice a common design problem concerns combined water versus graywater. Figure 3 shows the hydraulic capacity of sand filter columns loaded with different rates of graywater and combined water at a rate of 5 cm/day. The graywater had a COD equal to the combined-water, but a lower SS. The higher capacity in graywater columns was explained by the difference in SS. (It is uncertain what the "long term acceptance rate" will be with the two types of water since the experiment has only been run 1½ years with graywater.) It is usual to size an absorption field according to percolation rate and daily water usage. It seems, however, that a certain reduction in size may be allowed depending on the type of wastewater, but more field work will have to be done before such recommendations can be made. The difference in clogging rate in columns amended with graywater and combined-water was, however, not great in this experiment.

EFFECT OF SOIL TYPE

One of the most important factors influencing the physical part of the clogging process is the soil pore size distribution relative to the size distribution of the particles in the water (Behnke, 1969). As with the wastewater type, the effect of different soil types on clogging is rather confusing. Laak (1980) states that the biocrust is essentially the same for every soil, and that the permeability needs not to be measured for biocrusting design. EPA (1978) has, however, made a table for recommended maximum loading rates based on in situ measurements. The rates are from 1 cm/day for clays to 5 cm/day for sands. Loams have to be loaded at rates between these two values.

In one column experiment it is hoped to elucidate the effect of grain size and grain size distribution on clogging under different loading conditions.

THE EFFECT OF TEMPERATURE

As mentioned previously, some of the fascinating aspects of clogging research are all the contradictory results and opinions as to how the process is influenced by environmental conditions. Gupta & Swartzendruber (1962) in the earlier mentioned experiments with phenol additions found the same effects of low temperatures (1.5 ^{o}C) as with sterilizing agents, whereas Kristiansen (1981 a) found a higher clogging intensity in sand filters kept at 15-20 ^{o}C than at ambient temperatures. Increased clogging at low temperatures is, however, the most usual finding (De Vries, 1972; Simons & Magdoff, 1979, Stavn & Kristiansen, 1981).

A possible explanation of this is that most of the low temperature experiments are short-time experiments under highly artificial laboratory conditions. Temperature has probably different effects in different phases of clogging. Elevated temperature may have a minor effect on the initial accumulation of suspended solids or will only give a small increase in decomposition of SS. A higher temperature will also increase biological growth resulting in pore blockage in the last clogging phase. However, this needs to be confirmed by research.

THE EFFECT OF REDOX CONDITIONS

The extensive research at the University of California, Berkeley, in the sixties emphasized the importance of anaerobiosis on clogging (McGauhey & Winneberger, 1964). Since then many important studies have been undertaken to elucidate these effects.

Most of the investigations have been done from the point of view of dosing and resting versus continuous loading.

In a leaching field with a mature clogging mat, ponded water above the mat and constructed above ground water level, the lowest redox potentials will be found in the crust layer and higher potentials will occur underneath (Figure 2), because of aeration from the surrounding soil (Kristiansen, 1981 a). The question is whether this anaerobic crust layer has an unfavorable effect on the absorption capacity of the field. Would it be better to operate the field with resting periods in such a way that the crust becomes aerobic? In addition to the fact that it is a problem finding an optimum resting period for aerobic conditions in combination with an economical size of the septic system, studies over the last years have shown that infiltration capacity may be extended by maintaining an anaerobic environment (EPA, 1978; Kropf et al., 1975). Aerated columns were found to give increased clogging and this was explained by greater production of microbial slimes and/or biomass, or the formation of oxidized iron compounds directly below the black surface crust layer. Dosing and resting may therefore be a questionable manner of operating an absorption field unless the resting period is long enough to allow significant aerobic decomposition of the clogging mat.

The oxidation of organic material was studied in respiration cabinets filled with about 200 g of sand from a two year old clogging mat. An example of such an experiment is shown in Figure 4. As shown in the figure, it takes some hours' incubation in an aerobic environment before an effective microflora is developed or the right enzymes induced. First, the most easily usable substrate is broken down (the little peak) and thereafter a relatively constant respiration rate is established. After a period of about nine days about 20 percent of the total organic C was oxidized to CO_2 in this experiment. This is an example of

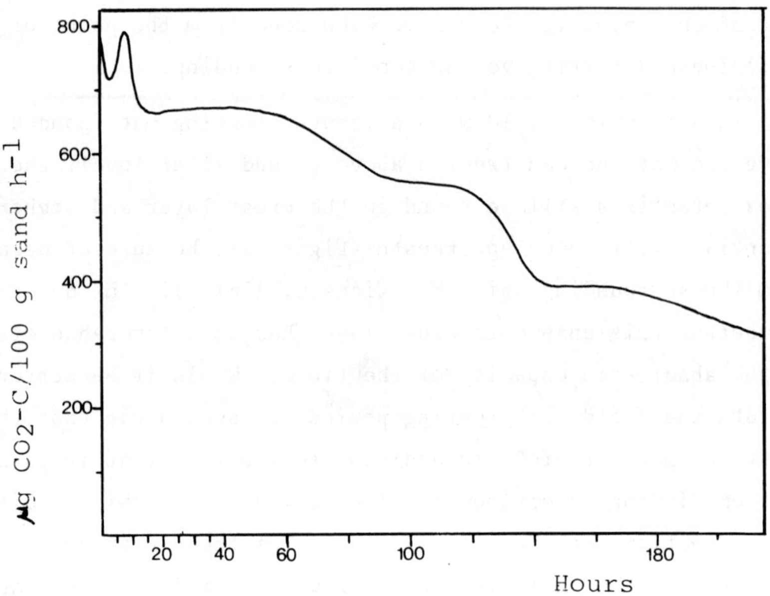

Figure 4. Rate of respiration of organic material from a two year old clogging layer in the laboratory.

clogging layer decomposition under optimum conditions in the laboratory at 22 °C with suction of plenty of air through the soil. In a septic system the temperature is significantly lower. Because of the considerable water holding capacity of the clogging layer it will take several weeks before the whole mat is aerobic. It was also probale that a portion of the remaining organic material after some days' respiration will be semi-stabilized "humic-like material" with a very slow turnover rate similar to that of native soil organic matter as shown with manure decomposition (Gilmour et al., 1977).

The low temperature in the field will also slow down the decomposition process (Figure 5). The resulting effect of resting on clogging mat removal must therefore be expected to be low. The author therefore recommends continuous loading of the field for as long as possible and resting periods of at least half a year.

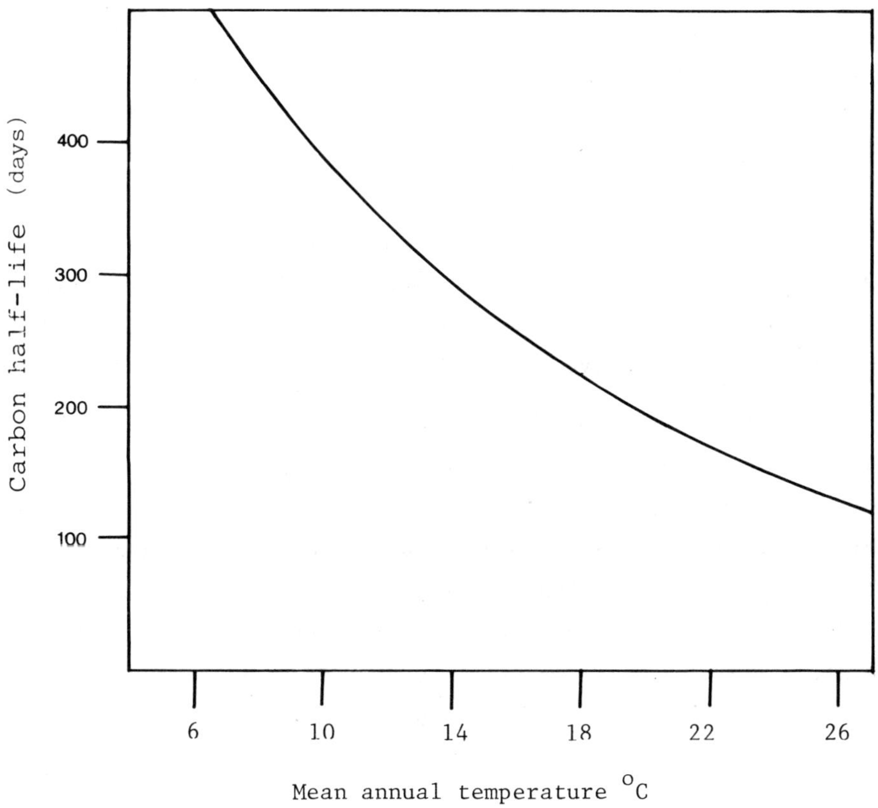

Figure 5. Effect of temperature on biodegradation of organic C calculated as carbon half-life.
(Data from Gilmour et al., 1975.)

THE EFFECTS OF CLOGGING

The effects of the biocrust layer on the hydraulic capacity of a leaching field is well known since Ryon's investigations more than fifty years ago. The design flow-rates are therefore 98-99 percent lower than measured field percolation rates.

In addition to the effect on hydraulic capacity, the effects on the renovation capacity are also dramatic as shown in the literature (EPA, 1978; Kristiansen, 1981 b, c; Laak, 1980). The crust layer restricts flow and induces a higher degree of unsaturation and thereby aeration (Figure 2) in the underlying soil. The degree of contact between the moving pollutant and the mineral particles will be increased with increasing clogging, resulting in a better renovation ability of the soil. The clogging zone also prevents the possibility of "short circuiting" (Tyler et al., 1977). The higher degree of unsaturation will also increase nitrification efficiency (Kristiansen, 1981 b; Sikora & Corey, 1976). Some of the same effects on degree of unsaturation and renovation capacities may be obtained by equal distribution on non clogged infiltrative surfaces.

The crust zone itself may be looked upon as a fisherman's net catching unwanted organisms before they penetrate into the soil. The largest amount of bacteria in the septic system is in this layer, and the microflora probably produces polysaccharides which take part in the "pathogen-catching". Such polysaccharides are beautifully described in relation to dental plaques (Costerton er al., 1978). Similar materials may be found to be the real clogging agent and the key to the understanding of biological clogging.

REFERENCES

Alexander, M.: 1961,"Introduction to Soil Microbiology". John Wiley & Sons, New York. 472 pp.

Allison, L.E.: 1947, "Effect of microorganisms on permeability of soil under prolonged submergence". Soil Sci. 63: pp. 439-450.

Avnimelech, Y. & Z. Nevo: 1964, "Biological clogging of sands". Soil Sci. 98: pp. 222-226.

Behnke, J.J.: 1969, "Clogging in surface spreading operations for artificial ground-water recharge". Water Resour. Res. 5: pp. 870-876.

Bouma, J., W.A. Ziebell, W.G. Walker, P.G. Olcott, E. McCoy & F.D. Hole: 1972, "Soil disposal of septic tank effluent: A field study of some major soils in Wisconsin". Information Circ. No. 20, Geol. Nat. Hist. Surv. Madison, USA. 240 pp.

Costerton, J.W., G.G. Geesy & K.J. Cheng: 1978, "How bacteria stick". Scientific American. January: pp. 86-95.

Daniel, T.C. & J. Bouma: 1974, "Column studies of soil clogging in a slowly permeable soil as a function of effluent quality". J. Environ. Qual. 3: pp. 321-326.

DeVries, J: 1972, "Soil filtration of wastewater effluent and the mechanism of pore clogging". J.W.P.C., 44: pp. 565-573.

EPA: 1978, "Management of small waste flows". Municipal Environ. Res. Lab. Office Res. Dev. Cincinnati, Ohio. EPA-600/2-8-173.

Gilmour, C.M., F.E. Broadbent & S.M. Beck: 1977, "Recycling of carbon and nitrogen through land disposal of various wastes". In: Soils for Management of Organic Wastes and Waste Waters. (ed. R.C. Dinauer). Publ. Soil Science of America. Madison, Wisconsin, USA. pp. 172-194.

Gupta, R.P. & D. Swartzendruber: 1962, "Flow-associated reduction in the hydraulic conductivity of quarts sand". Soil Sci. Soc. Amer. Proc. 26: pp. 6-10.

Hill, D.E. & C.R. Frink: 1980, "Septic system longevity increased by improved design". J.W.P.C. Aug., pp. 2199-2203.

Jones, J.H. & G.S. Taylor: 1965, "Septic tank effluent percolation through sands under laboratory conditions". Soil Sci. 99: pp. 301-309.

Kristiansen, R.: 1981 a, "Sand-filter trenches for purification of septic tank effluent. I. The clogging mechanism and soil physical environment". J. Environ. Qual. 10: pp. 353-357.

Kristiansen, R.: 1981 b, "Sand-filter trenches for purification of septic tank effluent. II. The fate of nitrogen."J. Environ. Qual. 10: pp. 358-361.

Kristiansen, R.: 1981 c, "Sand-filter trenches for purification of septic tank effluent. III. The microflora."J. Environ. Qual. 10: pp. 361-364.

Kropf, F.W., K.A. Healey & R. Laak: 1975, "Soil clogging in subsurface absorption systems for liquid domestic wastes". Progress in Water Technology 7: pp. 767-774.

Laak, R., K.A. Healey & D.M. Hardisty: 1974, "Rational basis for septic tank system design". Ground Water 12: pp. 348-351.

Laak, R.: 1980, "Wastewater Engineering Design for Unsewered Areas". Ann Arbor Science Publ. Michigan, USA. 179 pp.

Laak, R.: 1981, Personal communication.

Magdoff, F.R. & J. Bouma: 1974, "The development of soil clogging in sands leached with septic tank effluent". Proc. Home Sewage Disposal. Amer. Soc. Agric. Eng., pp. 37-47.

McGauhey, P.H. & J.H. Winneberger: 1964, "Studies of the failure of septic tank percolation systems". J. Water Pollut. Contr. Fed. 36: pp. 593-606.

Mitchell, R. & Z. Nevo: 1964, "Effect of bacterial polysaccharide accumulation on infiltration of water through sand". Appl. Microbiol. 12: pp. 219-233.

National Environmental Health Association: 1979:"1979 State of the Art Manual of On-site Wastewater Management". By the National Environmental Health Association's On-Site Wastewater Management Committee. Denver, Colorado, USA. 108 pp.

Nilsson, P.: 1981, Personal communication.

Sikora, L.J. & R.B. Corey: 1976, "Fate of nitrogen and phosphorus in soils under septic tank waste disposal fields". Trans. Amer. Soc. Agr. 19: pp. 866-875.

Simons, A.P. & F.R. Magdoff: 1979, "Laboratory evaluation of design parameters for mound system disposal of septic tank effluent". J. Environ. Qual. 8: pp. 486-492.

Stavn, A. & R. Kristiansen: 1981, "Gjentetting og omsetning i sandsøyler tilført avløpsvann". In: Forskningsprogrammet Rensing av avløpsvann i jord - spredt bosetning. Resultater fra forsøksarbeidet i 1980. Ås-NLH, pp. 19-37.

Thomas, R.E., W.A. Schwartz & T.W. Bendixen: 1966, "Soil chemical changes and infiltration rate reduction under sewage spreading". Soil Sci. Soc. Amer. Proc. 30: pp. 641-646.

Tyler, E.J., R. Laak, E. McCoy & S.S. Sandhu: 1977, "The soil as a treatment system". Paper presented at the Second National Home Sewage Treatment Symposium. Chicago, Illinois, USA. December.

Walker, W.G., J. Bouma, D.R. Keeney & P.G. Olcott: 1973, "Nitrogen transformations during subsurface disposal of septic tank effluent in sands. II. Ground water quality". J. Environ. Qual. 2: pp. 521-525.

THE SOIL AS A RENOVATING MEDIUM
THE FATE OF POLLUTANTS IN SOIL - ORGANIC MATERIAL

Rolv Kristiansen

Research Coordinator, Agricultural Research Council
of Norway, Section Soil Pollution Research,
1432 Ås - NLH, Norway

INTRODUCTION

The old system of disposing waste on people's own property is less common today. Water is now in many cases the transport medium of these wastes. This discharge of wastes into receiving waters has interfered with the ecological balance in such a way that many streams, lakes or estuaries are now deficient in oxygen. In addition, the esthetic values of the water have deteriorated. Consequently, organic matter and phosphorus removal have been the main goal of sewage treatment in Scandinavia.

The organic material in sewage comprises a variety of different compounds, each of them with a special composition that influences their biodegradability. The chemical composition of septic tank effluent has not yet been well documented. The composition will, however, vary from household to household depending on the habit of the occupants. Papers have been presented describing the types of organic material in primary and secondary effluent (Hunter & Kotalik, 1973; Painter et al., 1961; Rebhun & Manka, 1971). Painter's group succeeded in identifying 79 percent of the organic carbon in settled sewage into different

A. S. Eikum and R. W. Seabloom (eds.), Alternative Wastewater Treatment, 121–128.

chemical groups. The remaining 20 percent which they were not able to identify, probably consisted of complex organic material like humic acids (Rebhun & Manka, 1971). About half the identified carbon in solution was in the form of carbohydrates, and fats made up the largest portion of suspended solids. Anaerobic treatment brought about decreases in soluble carbohydrates (80-95 percent), and the concentration of acids especially acetic acid increased. The organic material in septic tank effluent is probably similar to the settled sewage referred to above.

The large number of different types of organic material in sewage, variations between households, and the practical problems met when analyzing, make BOD, COD, TOD, and TOC the usual way of characterizing the amount of organic matter in different types of sewage effluent. The difference between COD and BOD is often used as a measure of the amount of recalcitrant molecules.

In the USA, BOD and COD from combined septic tank effluent are, respectively, reported to be 100-250 and 200-300 mg 0/1 (EPA, 1978) while somewhat higher values are reported in Scandinavia (Lindbak, 1978; SNV, 1980). In graywater the author has found COD to be 300-400 mg 0/1 while typical BOD values were between 100 and 200 mg 0/1. A further discussion of the composition of various types of septic tank effluent has been given by Professor Boyle during this conference.

Generally, soil filters are known for their effective degradation of organic material and therefore often mentioned as "living filters". In the writer's opinion a description of organic matter removal must be linked to a short introduction to the microflora of such types of biological filters since it is this flora that produces the enzymes necessary for biochemical degradation and mineralization.

THE MICROFLORA OF SEPTIC SYSTEMS

All who are engaged in the agricultural sciences are well aware of the potential of soil microflora in organic matter degradation, and textbooks on soil microbiology have been published (Alexander, 1961). When soil is amended with septic tank effluent, the types of the natural soil and water flora that are most fitted for the changed soil environment, will develop while other types will not proliferate at a comparable speed.

Figure 1 shows plate counts from three different depths in the overloaded end of a standard sand-filter trench loaded with about 5 cm septic tank effluent per day. The first point is the average background value for this specific sand. Loading of the sand-filter trench with septic tank effluent resulted in an increased number of bacteria in the sand fill material. The number of viable bacteria (estimated by plate counts on nutrient agar) in the sand surface increased from an initial number of 10^6 to more than 10^8/g dry sand. Below the sand surface, the number of viable bacteria in the fill was about 10^6/g dry sand.

Generally, 10 to 90 percent of the count at 20 °C was found when using an incubation temperature of 37 °C. The highest percentage was found in the summer. As shown in Figure 1, the number of proteolytic bacteria increased on the whole in the sand fill, but only amounted to a few percent of the total amount of viable bacteria in the sand. There was a higher percentage of lipolytic bacteria in the higher soil horizons which indicates that most of the fat is broken down in those zones of the filter. The absolute numbers of streptomycetes were either the same as in unloaded sand (about 5×10^4/g) or decreased to about 10 percent of this value. The fact that streptomycete hyphae were seldom detected in suspensions of sand studied under the light microscope, supports the results from the plate counts and suggests that few actively growing streptomycetes occur in this type of

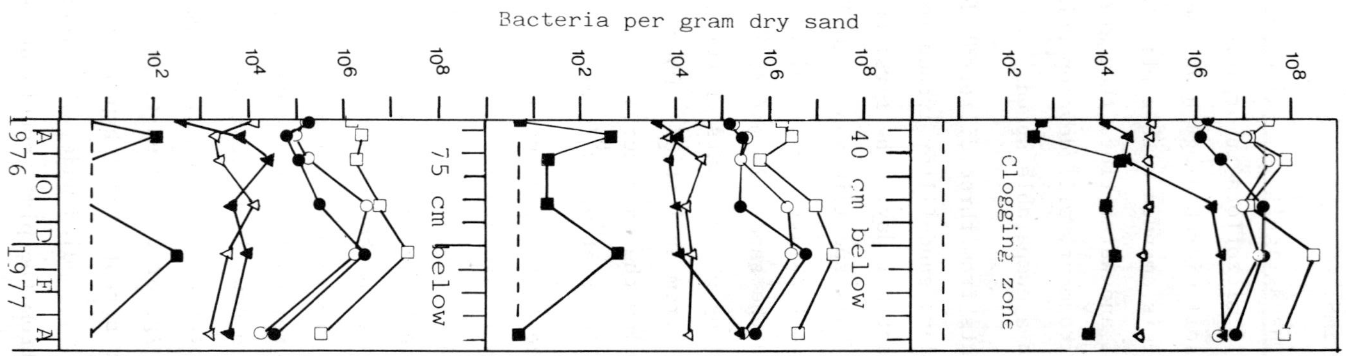

Figure 1. Plate counts from three different depths in the overloaded end of a sand-filter trench loaded with 5 cm septic tank effluent per day.

Symbols: Detection limit (---) 5 cells/g sand; viable count on nutrient agar incubated at 20 °C (□) and 37 °C (○); fecal coliforms (■); proteolytic bacteria (●); lipolytic bacteria (▼); and actinomycetes (▽).

septic system. There is, however, no reason for expecting them to be more important in absorption beds laid in natural soil. The total amount of bacteria in sand-filter trenches is about 100 times as high as the viable count (Kristiansen, 1981 b) and the clogging intensity also influences the number and morphological type of bacteria.

Soil absorption beds have confined anaerobic/aerobic zones in addition to possible anaerobic microzones in apparently aerobic soil (Kristiansen, 1981 a). A certain amount of the complex molecules are degraded anaerobically in the clogging layer and by-products are leaked to aerobic zones below where they are mineralized. The energy output from an aerobic metabolism is much higher than from an anaerobic metabolism. The biomass which may be developed on a certain amount of organic material, is therefore highest in interfaces between the anaerobic clogging layer and the aerobic zone below where acids from the clogging layer are mineralized (Kristiansen, 1981 b). Investigation of the microflora have also shown that a large proportion of the bacteria in the clogging layer are of "effluent origin" and do not take a significant part in the degradation process while most of the bacteria in aerobic zones are active in the decomposition of organic material in septic tank effluent (Kristiansen, 1981 b).

DECOMPOSITION OF ORGANIC MATERIAL

The potential for organic matter removal in soil filters depends on the integrated effect of factors such as type of organic material, size of surface area, degree of aerobicity, loading rate, temperature etc. Generally, all sorts of biological filters are considered to be good BOD removers.

In column investigations with trickling filters having a specific area 240 m^2/m^3, a stabile and low content of organic material was found in the column effluent when the loading rate was kept below 160-200 $l/m^2/h$ (Brattebø, 1980). The COD removal was more than 90 percent, independent of COD in inlet water in the range of 200-600 mg O/l. It was, however, necessary to remove sludge from the filter effluent in order to maintain a constant low level of SS.

Columns (75x15 cm) filled with different sizes of gravel (4-8 mm and 8-15 mm), loaded with 5 cm septic tank effluent in a room at 12 ^{o}C showed a SS removal of 60-80 percent, independent of gravel size, and a COD removal of 60-80 percent and 40-60 percent for the two types of gravel (Stavn & Kristiansen, 1981). At certain times sludge escaped from the columns. Long term efficiency is therefore dependent on sludge removal. Gravel filters must be expected to be sensitive to low temperatures. EPA (1978) considers the large head requirements for deep excavation or pumping above ground to be disadvantages with on-site trickling filters. Some manufacturers of commercial filters have, however, solved these problems.

Sand filter columns run at 4, 8 and 12 ^{o}C at a rate of 5 cm/day removed 60-80 percent of COD (Stavn & Kristiansen, 1981). There were insignificant differences between columns run at 8 and 12 ^{o}C while the 4 ^{o}C columns always showed a low efficiency during the 10 month experimental period. COD in effluent from the low temperature columns were higher than 50 mg O/l. Sand filter columns run in a study in Sweden did not show any temperature effects between 8 and 25 ^{o}C (VIAK, 1979).

Elevated TOC concentrations were not found to have any effect on TOC removal efficiency. A consistent removal of about 90 percent organic matter from both combined- and graywater by soil systems is also reported by EPA (1978). Siegrist (1980) reports only 50 percent COD removal in graywater sand filters loaded at a rate of 30 cm/day. It was felt that this was due to the combined effect of the high loading rate and low temperature in his filters.

If one calculates BOD/COD from existing sand-filter data (Lindbak, 1978), the values for septic tank effluent will be found to be 0.6-0.7 and for filter effluent 0.2-0.3. This indicates that organic matter in outlet water from soil systems contains a considerable proportion of recalcitrant organic materials like humic acids. This may also be shown by following the fate of nitrogen (Kristiansen, 1981 b).

Generally, organic matter removal is not considered a problem in septic systems if loaded as recommended to maintain the hydraulic capacity and removal of other pollutants. Examinations of field systems both in this country and others have confirmed this.

REFERENCES

Alexander, M.: 1961, "Introduction to Soil Microbiology". John Wiley & Sons, New York. 472 pp.

Bratteb⌀, H.: 1981, Rapport fra prosjektet "Rensing av avl⌀psvann for spredt bebyggelse". Institutt for vassbygging - NTH, Trondheim.

EPA: 1978, "Management of small waste flows". Municipal Environ. Res. Lab. Office Res. Dev. Cincinnati, Ohio. EPA-600/2-8-173.

Hunter, J.V. & T.A. Kotalik: 1973, "Treated municipal wastewater - what is it? 1. Chemical and biological quality of treated sewage effluents". In: Sopper, W.E. & L.T. Kardos (ed): Recycling Treated Municipal Wastewater and Sludge through Forest and Cropland. The Penn. State Univer. Press, Pennsylvania, pp. 6-25.

Kristiansen, R.:1981 a, "Sand-filter trenches for purification of septic tank effluent. I. The clogging mechanism and soil physical environment". J. Environ. Qual. 10: pp. 353-357.

Kristiansen, R.: 1981 b, "Sand-filter trenches for purification of septic tank effluent. III. The microflora". J. Environ. Qual. 10: pp. 358-361.

Lindbak, P.: 1978, "Rensing av avløpsvann fra spredt bolig- og fritidsbebyggelse. PRA 20, Oslo. 54 pp.

Painter, H.A., M. Viney & A. Bywaters: 1961, "Composition of sewage and sewage effluents. J. Proc. Inst. Sew. Purif. pp. 302-314.

Rebhun, M. & J. Manka: 1971, "Classification of organics in secondary effluent". Environ. Sci. Tech. 5: pp. 606-609.

Siegrist, R.L.: 1980, "Graywater treatment by coarse media filtration of septic tank effluent." Small Scale Waste Management Project. University of Wisconsin, Madison. Paper presented at the 3rd Northwest On-Site Waste Water Disposal Short Course, University of Washington, Seattle, March 4-5.

SNV: 1980, Statens Naturvårdsverk, "Infiltration av husholdsavloppsvatten". SNV Meddelande 4/1980, 52 pp.

Stavn, A. & R. Kristiansen: 1981, "Gjentetting og omsetning i sand-søyler tilført avløpsvann".
In: Rapport nr. 2 fra forskningsprogrammet "Rensing av avløpsvann i jord - spredt bosetning. pp. 18-37.

VIAK: 1979, "Avloppsinfiltration". Slutrapport från forskningsprojektet Avloppsinfiltration - Grundvattenpåverkan, av Kaj Nilsson & Peter Englöv, 193 pp.

ON-SITE SOIL SYSTEMS, NITROGEN REMOVAL

Rein Laak

Professor of Civil Engineering,
University of Connecticut, Storrs,
Connecticut 06268, USA

INTRODUCTION

The problems resulting from uncontrolled discharge of nitrogen-rich wastewater range from eutrophication in receiving waters to methemoglobinemia and interference with the cardiac function in humans (1). These and other effects have been evaluated and documented by a number of workers (2, 3). In the design of on-site wastewater treatment facilities, one is therefore responsible for the control of nitrogen release in order to protect ground and surface water resources. A biological nitrogen removal technique which is passive, energy free and requires no more maintenance than a conventional septic tank system would be useful for areas where nitrate pollution from septic tank systems may be a potential problem.

A review of the fate of nitrogen in septic tank and subsurface fields will be presented. A passive nitrogen removal technique developed at the University of Connecticut, called the RUCK system, will be discussed.

A. S. Eikum and R. W. Seabloom (eds.), Alternative Wastewater Treatment, 129–143.

SEPTIC TANK AND SOIL SYSTEM

Total Kjeldahl Nitrogen (TKN) in domestic sewage is taken to be the summation of ammonia, urea, amino acids and proteins, but the single largest source is urea, the precursor and the major fraction of ammonia. Raw sewage contains about 40 mg/l TKN and insignificant amounts of nitrite and nitrate. The major source of TKN is urine; about 80 percent of the total nitrogen from households (4, 6). The average mass of nitrogen wasted per capita per day is about 12 g (4). The amount of urea excreted per day per capita is about 35 g (7).

The septic tank accumulates septage or sludge which contains nitrogen (6, 8). About 10 percent of the total nitrogen in raw sewage is removed via sludge in a septic tank (9). The septic tank converts some of the organic nitrogen to ammonia, resulting in an effluent containing about 70 percent ammonia and 30 percent organic nitrogen (5, 6).

At the soil interface, the clogging layer or mat enhances hydrolysis of urea and further ammonification. The nitrogen, upon entering the soil, may undergo further ammonification, nitrification, adsorption, ion exchange, fixation, volatilization, biological uptake, and/or eventual denitrification.

a) Ammonification

The biodegradable portion of the organic nitrogen is decomposed by many soil bacteria releasing ammonium. The reaction can occur under either aerobic or anaerobic conditions. The simplified reaction may be represented by the equation (10, 11):

$$\text{Organic N} \xrightarrow{\text{microorganisms}} \frac{NH_3}{NH_4^+} + \text{other products} \qquad \text{(I)}$$

b) Nitrification

Nitrification is an aerobic biological reaction that occurs in at least two steps to ultimately form nitrate. The first reaction is carried out by nitrobacter and produces nitrite. A simplified equation is (12):

$$NH_4^+ + 1\tfrac{1}{2}\ O_2 \xrightarrow{\text{nitrobacter}} NO_2^- + 2H^+ + H_2O + \text{energy} \qquad \text{(II)}$$

The second reaction is accomplished by nitrosomonas and to a lesser extent by nitrosococcus, nitrospira, nitrosocystis and nitrosogloes (13). The simplified equation for this reaction can be represented by (12):

$$NO_2^- + \tfrac{1}{2}\ O_2 \xrightarrow{\text{nitrosomonas}} NO_3^- + \text{energy} \qquad \text{(III)}$$

The bacteria affecting this reaction are referred to as chemoautotropic bacteria. Carbon dioxide is used as a carbon source for production of new cell material. The energy for the synthesis is provided by oxidation of inorganic substrate, i.e., ammonium.

Nitrification occurs very commonly in the zone of aeration between the seepage trench clog layer and capillary zone. Under warm, aerobic soil (pH 5.6-8.0) conditions a few inches of soil thickness will convert ammonia to nitrate. It fails to occur in systems where the ground water is very near to the bottom of the leaching trench so that anaerobic conditions prevail (14). The public health and aesthetic significance of the nitrate ion coupled with the ease of migration of the nitrate ion in soils make the nitrification reaction highly important. Danger to potable well water supplies can sometimes exist in intensely populated areas where both leaching fields and well water supplies are used in well drained sand gravel soils.

c) Adsorption

In cases where nitrification does not predominate, adsorption can be significant. Only in conditions of insufficient aeration or absence of proper bacteria does nitrification not predominate. Insufficient aeration exists when the ground water table rises to near the bottom of a leaching trench or higher or in seepage through slowly permeable soils (14, 15). A second condition for significant adsorption of ammonium to occur is the presence of sufficient quantities of negatively charged clay and organic colloids, almost the only determinant of adsorption capacity. Adsorption capacity may vary from 2 mg N/100 g soil for a sandy soil to 100 mg N/100 g soil for a fine grained soil (30 percent clay) (16, 17).

Physical adsorption of ammonium occurs either by hydrogen bonding with an oxygen on the exchange complex of a colloid or by hydrate formation with salts on the exchange complex (e.g. $CaCl_2 \cdot 2NH_3$). Chemical adsorption occurs when ammonia reacts with a hydrogen atom in the exchange complex forming ammonium which becomes the exchangeable ion on the exchange complex. Physical adsorption is characterized by having greater reversibility and lower energy (18, 19).

Adsorption is reversible and subject to nitrification if reaeration occurs and proper bacteria are established or to plant uptake. Adsorption may detain nitrogen in soil for subsequent loss through plant uptake. The nitrogen also can be only temporarily concentrated until conditions become favorable for nitrification in which case high concentrations of the nitrate ion can be leached into ground water. This can happen with the rise and fall of water tables under leaching trenches (11).

Fixation in contrast to adsorption is relatively stable and resists nitrification and plant uptake. Fixation can be by the clay and/or the organic fraction of the soil. It occurs in clay by entrapment of the ammonium ion between the intermicellular layers of clay minerals such as montmorillonite or vermiculite. Ammonia can be fixed by the organic portion of soil by forming very stable complexes. In some soils the organic fraction may play a greater role than the mineral fraction (11, 18).

d) Ammonia Volatilization

The equilibrium concentrations of ammonium (NH_4^+) and volatile ammonia (NH_3) are a function of the pH (20). It can be seen that at the pH values typical of domestic wastewater (6.5 to 7.5) very little ammonia is present. In addition a high air-water contact would be necessary to allow escape of the ammonia (21). This is not provided in a subsurface seepage field. For these reasons, therefore, little volatilization is expected (11, 14).

e) Biological Uptake

Plants can remove significant amounts of nitrogen from soils. In land applications wastewater filters through the plant root zone. Removals of 100-200 kg ha/yr and as high as 500 kg ha/yr are reported depending in the crop (21). Wastewater from subsurface seepage field, however, is not as available to plants, although some nitrogen can be removed if roots are deep (22).

Nitrogen is a necessary nutrient for bacteria and other micro-organisms. Upon death and cell decomposition the nitrogen is at least partially released. Research has indicated, however, that nitrogen incorporated into microbe tissue is held in a rather stable form (23, 24).

f) Denitrification

Denitrification, when it occurs, closes the loop of the nitrogen cycle by reducing nitrates to inert N_2 gas. The reaction is represented by the following equation (11, 19):

$$NO_3^- + \text{carbon source} \xrightarrow[\text{bacteria}]{\text{denitrifying}} N_2 + H_2O + CO_2 + \text{cells} \qquad \text{(IV)}$$

The bacteria use the nitrate as an electron acceptor in anaerobic environments. Temperature and pH should be above 10 ^{o}C and 5.5 respectively for maximum reaction rates (11). The reaction can be carried out by Pseudomonas, Achromobacter, Bacillus and Micrococcus (13).

Much of the biodegradable carbonaceous material is stabilized after filtration through only one or two feet. For denitrification to occur, nitrates must pass into an anaerobic environment along with sufficient carbon source to support the denitrification reation. Winneberger (25) argues that significant denitrification occurs in soils in zones of "micro-anaerobiosis" such as around decaying roots. Other investigators (26, 27) have been able to create the conditions for soil denitrification in laboratory setups.

g) Leaching Fields

In leaching fields (one meter of unsaturated flow in aerobic soil between the seepage trench and the ground water) nitrification followed and leaching of the nitrate into the ground water occurred (14, 15, 28). Dilution is the main mechanism available to reduce nitrate concentrations to safe levels. In conditions of high ground water or very slowly permeable soils (anaerobic conditions) adsorption of ammonia onto the clay and organic fraction of soils occurs (14, 17). Ammonium travels as adsorption sites become exhausted.(22). Most of the ammonium is subject to nitrification and leaching if aerobic conditions are reestablished (11). One half meter of concrete sand with a mature clogging layer was found to remove about 10 percent of total nitrogen (9).

DESCRIPTION OF THE RUCK SYSTEM

The typical biological nitrogen removal techniques have been described (4, 6). The RUCK system separates plumbing drain lines into two waste streams. The nitrogen rich stream (the toilet wastes) called blackwater is treated using a septic tank followed by an underdrained aerobic sand filter. The underdrained sand filter nitrifies the nitrogen. All or part of the remainder of wastewater, called graywater, contains an abundance of organic carbon and is treated in a separate septic tank to remove settleable solids. If garbage grinders could be used, a significant increase in the graywater organic carbon would be achieved.

The two waste streams are brought together in an anaerobic upflow rock-filled tank where biological denitrification occurs. The rock filter effluent is disposed in conventional seepage trenches. The home system operates by gravity like a standard septic tank system.

Experimental Procedure

Before actual implementation of a full scale home unit model, two laboratory studies were performed.

The first laboratory study to determine the applicability of graywater in the denitrification step of the process consisted of measuring carbon degradation rates with a respirometer containing suspended cells (30). A variety of volume ratios of methanol, graywater and raw settled sewage were injected into the 374 ml aparatus.

The second laboratory study utilized six nitrifying sand-column reactors followed by six fixed media reactors, each pair receiving either methanol, graywater or settled sewage as a carbon source. A 0.1 m diameter, 1.52 m high plexiglass column packed to a depth of 0.61 m with concrete sand comprised the nitrifying packed-column reactors. The sand used in the columns had a uniformity coefficient of 4.0 and an effective size of 0.16 mm.

The fixed media columns consisted of 0.1 m diameter, 1.52 m high plexiglass columns filled to a depth of 0.61 m with 0.05 to

0.1 m stones. Provision was made for the addition of the three organic carbon solutions (220 ml/reactor/day) prior to the denitrification step in the rock filled columns.

Conventional septic tank effluent taken from the University of Connecticut Wastewater Experiment Station was dosed two times per day into the nitrifying sand-column reactors at a rate of 6 cm/d. The upflow rock filled reactors were connected to the sand-column reactors in a fashion ensuring anaerobiosis. Sampling ports were provided and samples were collected five times per week.

Analyses for nitrate, nitrite, and ammonia-nitrogen were determined with a Technicon Auto Analyzer. BOD_5, suspended solids, dissolved oxygen, pH, TKN, and temperature were determined as per Standard Methods (31). Due to budget limitations our analysis work was kept to a minimum.

A field demonstration model for a home was designed on the basis of the following criteria.

a) Blackwater contains most of the household wastewater nitrogen (over 80 percent) and represents 40 percent of the total flow. The existing 2 m^3 septic tank could be used for the blackwater pretreatment. The subsurface filter bed sand should not be finer than ES = 0.16 mm U.C. $\leq$ 5 if loaded over 3 cm/d. The nitrification step requires alkalinity/nitrogen = 7/1 and an airflow of 1.5 m^3/m^3 wastewater. The sand filter area needed was 12 m^2 and 0.5 m deep to to process 40 percent of the total flow.

b) Graywater represents 60 percent of the household flow and contains sufficient organic carbon for denitrification. The carbon in graywater biodegrades faster than blackwater carbon. The graywater septic tank should not be any larger than 2 m^3 to prevent loss of organic carbon by over-treatment.

c) The anoxic 3.8 m^3 denitrification reactor was based upon five day liquid detention time with no deliberate cell wastage. The specific fixed surface area was estimated to be 600 m^2 using 0.05 m rocks.

d) The RUCK system appeared to have advantages. The graywater septic tank would accumulate solids about eight times less than a blackwater septic tank and therefore would need pumping once in 10 or 15 years. The blackwater tank carrying 40 percent of the liquid flow could be used to store a greater volume of sludge. The blackwater tank, if it were a conventional 3.8 m^3 tank, could have the added sludge storage space and reduced required liquid space. The septage pumped would contain less liquid and therefore the total volume of septage could be somewhat reduced.

The final effluent from the RUCK system would have received greater treatment and would impose a reduced clogging load on the seepage field soil interface. The final seepage field interface could be reduced or greater protection against failure could be achieved.

Results

The three experiments showed that the nitrification-denitrification process adequately removed nitrogen and that graywater was a suitable carbon source in the denitrification step.

Laboratory Study

a) Graywater as an organic carbon source

Results of the first laboratory study showed that graywater from a typical home contained abundant soluble carbon which biodegraded at the same rate as methanol. Oxygen uptake rate ratios for the suspended cells were calculated as follows:

graywater/raw sewage
methanol/raw sewage

The overall ratio was one.

b) Nitrification-denitrification process

The results of this study supported the previous findings and provided additional information as a basis for design criteria.

Nitrification: Ammonia oxidation and BOD_5 removals in the nitrifying reactors were essentially identical for all reactors after the initial acclimatization period. However, only limited nitrification and BOD_5 removals occurred in the reactor units studied. Since effluent dissolved oxygen concentrations averaged 2.8 mg/l in the aerobic reactors throughout this period of operation, it did not appear that the oxygen content would curtail nitrification. Only after the hydraulic loading rate was decreased from 20 cm/d to 6 cm/d, did a significant degree of nitrification and BOD_5 removal occur.

Suspended solids' concentrations ranged from 43 mg/l in the feed solution to 19 mg/l in the effluent of the nitrifying reactors. The pH in the system ranged from 6.8 to 7.2.

Increases in the BOD_5 concentrations were measured in the effluents of the denitrifying reactors using all three carbon sources.

Slightly higher suspended solids' concentrations did occur in the denitrification reactors' effluent. However, the slight addition in suspended solids after the denitrifying step was due to synthesized cellular microbial material being washed from the fixed-media reactors. The bacterial growth on the fixed-media remained approximately 1 mm thick and at no time did clogging of the fixed-media reactors occur. In comparing results from the reactors using the three investigated carbon sources, methanol contributed the least suspended solids to the effluent.

Nitrate reduction was measured in all denitrifying reactors receiving the three selected carbon sources. Reactors receiving graywater as a carbon source removed 70 percent of the available nitrate while methanol removed 40 percent and settled sewage removed less than 5 percent respectively. The low level of nitrate reduction was obtained with low loading rates of carbon source. Increasing the concentration of methanol added, resulted in 83 percent reduction in nitrate.

Field Study

The field model built in the fall 1977 had several problems which affected the early results. The blackwater septic tank in the spring of 1979 unloaded solids on the sand filter and nitrification dropped to 30 to 50 percent of the possible maximum. Prior to the spring of 1979 the filter was covered with 0.3 m of silty loam and nitrification was 50 to 80 percent of the possible maximum. Only after the ponded filter was unloaded and air vents were constructed, nitrification increased to near maximum values. The ammonium concentrations in the sand filter effluent dropped to less than 2 mg/l and the pH dropped to 3.6 to 4.0. The TKN values of the sand filter effluent were now less than 4 mg/l. In the fall of 1979 a second 2 m^3 septic tank was added in series to delay any immediate future solids' outwash from the blackwater septic tank. A second major problem developed. The graywater septic tank leaked at the horizontal joint at halfway depth for two years without sealing. A pump was necessary to lift the graywater. Various pumping rates were tried to balance the water level in the graywater tank. The rock filter leakages were also high. The net effect was that the rock filter did not fill up until fall 1978 although the soil had a percolation rate of 8 minutes per cm and a saturated permeability of about $5x10^{-4}$cm/s. The graywater tank was removed in the fall of 1979 and was replaced with a waterproof plastic septic tank. The final effluent from the RUCK system since replacement had a mean nitrate content of less than 7 mg/l (at 6 ^{o}C) and a TKN of less than 4 mg/l.

The nitrification process could be limited by available alkalinity in the blackwater, presumably because the nitrifying bacteria utilize inorganic carbon for all synthesis rather than available organic carbon. According to the values reported in literature (32) the alkalinity/nitrogen requirement of 7/1 could be met by blackwater. At the home system, alkalinity in the blackwater nitrification step was completely exhausted and pH dropped to 3.5-4.0. The well water alkalinity was measured to be about 40 mg/l as $CaCO_3$ which agreed closely with the mean alkalinity values for crystalline bedrock areas (32). Typical total alkalinity pick up from domestic water use (34) was reported to be 100-150 mg/l; however, evidence suggests that blackwater picks up significantly more alkalinity than graywater. In order for the RUCK system to remain passive, addition of alkalinity was not considered.

The unique feature of the RUCK system is the use of some or all of the graywater as a carbon source. It is comparable to methanol in effectiveness yet it constitutes no additional maintenance cost, is generated on-site and is relatively non-toxic. The system is an economical (estimated US$ 1,000 more than a standard septic tank system) and effective means of removing nitrogen from domestic wastewater.

CONCLUSIONS

In sandy soils where aerobic conditions are maintained, domestic septic tank effluent will be nitrified. Nitrate can move readily with water through soil. Ground water can be polluted with nitrate if sufficient dilution is not available from ground water flow. A passive denitrification system called RUCK has been investigated. The RUCK system utilized graywater as the organic carbon source for the denitrification step.

ACKNOWLEDGEMENTS

The RUCK system work was supported by the U.S. Department of Interior, Office of Water Research and Technology as authorized under the Water Resources Act of 1964, P.L. 88-379 as amended.

REFERENCES

1. "Accumulation of Nitrate". 1972, Committee on Nitrate Accumulation, Agricultural Board Division of Biology and Agriculture, National Research Council. National Academy of Sciences, Washington, D.C.

2. "Nitrogenous Compounds in the Environment". 1973, Hazardous Material Advisory Committee, EPA-SAB-73-001, U.S. Environmental Protection Agency, Washington, D.C.

3. Sawyer, C.N.: 1973, "Sources of Nitrogen in the Water Resource and Its Role in Eutrophication". Conference on Nitrogen Removal from Municipal Wastewater, University of Massachusetts, Water Resources Research Center.

4. "Process Design Manual for Nitrogen Control". 1975, U.S. Environmental Protection Agency, Technology Transfer, Washington, D.C.

5. Laak, R.: 1980, "Wastewater Engineering Design for Unsewered Areas". Ann Arbor Science.

6. EPA: 1980, "On-site Wastewater Treatment and Disposal Systems, Design Manual". U.S. EPA Technology Transfer.

7. Hanson, A.M. & T.F. Flynn: 1964, "Nitrogen Compounds in Sewage". Proc. 19th Ind. Waste Conf., Purdue Univ. Eng. Ext., No. 117, pp. 32-44.

8. Kolega, J.J. et al.: 1976, "Treatment and Disposal of Wastes Pumped from Septic Tanks". Storrs Agr. Exp. Sta., Univ. of Connecticut.

9. Hardisty, D.M.: 1974, "Nitrogen and Phosphorus Considerations in On-Site Wastewater Disposal". M.S. Thesis, Univ. of Conn.

10. Hendricks, D.W. & W.D. Pote: 1974, "Thermodynamic Analysis of a Primary Oxidation Pond". Journal of the Water Pollution Control Federation 46: pp 333, Feb.

11. Lance, J.C.: 1972, "Nitrogen Removal by Soil Mechanisms". Journal of the Water Pollution Control Federation 44: pp. 1352, July.

12. Fair, G.M., J.C. Geyer & D.A. Okun: 1971, Elements of Water Supply and Wastewater Disposal, John Wiley & Sons, New York, pp. 489, 535 and 712.

13. Reeves, T.G.: 1972, "Nitrogen Removal: A Literature Review". Journal of the Water Pollution Control Federation, 44:1895, Oct.

14. Dudley, J.G. & D.A. Stephenson: 1973, Nutrient Enrichment of Ground Water from Septic Tank Disposal Systems, University of Wisconsin, Madison, Wisconsin, Nov.

15. Preul, H.C.: 1967, "Underground Movement of Nitrogen". Proceedings Third International Conference Advances in Water Pollution Research, Munich, Germany, pp. 309.

16. Polta, R.C.: 1969, "Septic Tank Effluents". Water Pollution by Nutrients-Sources, Effects and Control, presented at Annual Meeting of Minnesota Chapter Soil Conservation Society of America, Univ. of Minnesota, Water Resources Research Center Bulletin 13.

17. Pruel, H.C. & G.J. Schroepfer: 1968, "Travel of Nitrogen in Soils". Journal of the Water Pollution Control Federation, 40:30, Jan.

18. Baily, G.W.: 1968, Role of Soils and Sediment in Water Pollution Control Part I Reactions of Nitrogenous and Phosphatic Compounds with Soils and Geologic Strata, U.S. Department of the Interior, Federal Water Pollution Control Administration, Southeast Water Laboratory, U.S. Government Non-Depository, W69-03080, Feb. 1973, 17936.

19. Tilstra, J.R., K.W. Malueg & W.L. Larsen: 1972, "Removal of Phosphorus and Nitrogen from Wastewater Effluents by Induced Soil Percolation". Journal of the Water Pollution Control Federation 44:796, May.

20. Blackburn, T.R.: 1969, Equilibrium, Holt, Rinehart and Winston, New York.

21. Pound, C.E. & R.W. Crites: 1973, Wastewater Treatment and Reuse by Land Application, Volume 2, Environmental Protection Agency, Office of Research and Development, Aug.

22. Soil Conservation Society of America, 1976, Land Application of Waste Materials.

23. Bartholomew, W.V.: 1965, "Mineralization and Immobilization of Nitrogen in the Decomposition of Plant and Animal Residues". In: "Soil Nitrogen", W.V. Bartholomew and F.E. Clark (Eds.), American Society of Agronomy, Inc., Madison, Wisconsin.

24. Frederick, L.R. & F.E. Broadbent: 1967, "Biological Interactions". In: "Agricultural Anhydrous Ammonia Technology and Use", M.A. McVicker et al., (Eds.), American Society Agronomy, Inc., Madison, Wisconsin.

25. Winneberger, J.H.: 1971, On-Site Waste Management, Vol. 2, Hancor, Inc.

26. Glandon, L.R. & L.A. Beck: 1969, "Agricultural Nitrate Reduction at a Water Table". Collected Papers Regarding Nitrates in Agricultural Wastewater, Water Pollution Control Federation Research Series, Dec. pp. 41-52.

27. Patrick, W.H. & M.E. Tusneem: 1972, "Nitrogen Loss from Flooded Soil". Ecology 53:735, Summer.

28. Bouma, J., W.A. Ziebell, W.G. Walker, P.G. Olcott, E. McCoy & F.D. Hole: 1972, Soil Absorption of Septic Tank Effluent, Information Circular Number 20, Univ. of Wisconsin.

29. Winneberger, J.H.T.: 1975, Manual of Gray Water Treatment Practice, Part II, Monogram Industries, Inc. California.

30. Anon., Wasserversorgung, Abwasserreinigung und Gewasserschutz I, Lecture Notes, Technical University, Vienna, Austria.

31. Standard Methods for Examination of Water and Wastewater, 13th ed., 1971. American Public Health Association, American Waterworks Association and Water Pollution Control Federation, Washington, D.C.

32. Brandes, M.: 1977, "Characteristics of Effluents from Separate Septic Tanks Treating Gray Water and Black Water from the Same Home". Ministry of the Environment, Ontario, Canada.

33. Thomas, M.P. et al.: 1967, "Water Resources Inventory of Connecticut". Part 2, Shetucket River Basin, prepared by USGS, Connecticut Water Resources Bulletin No. 11.

34. Metcalf and Eddy Inc.: 1979, "Wastewater Engineering: Treatment, Disposal, Reuse". 2nd Ed., McGraw-Hill Book Co.

PHOSPHORUS SORPTION BY SOIL; A REVIEW

Arne O. Stuanes

Researcher, Dr. Scient., Norwegian Forest Research Institute, The Agricultural University of Norway, P.O.Box 61, 1432 Ås-NLH, Norway

ABSTRACT

The soil factors affecting phosphorus sorption are discussed together with phosphate sorption isotherms and sorption mechanisms. Both soil column experiments and field studies have shown that suited soils have a high sorption capacity for phosphorus from wastewater. It is very important that a simple and accurate sorption index for phosphorus in soil and the treatment systems' operation time be found.

INTRODUCTION

Phosphorus is often the limiting nutrient for plant growth in watercourses, and increased input can therefore cause eutrophication. All possible efforts should be made to clean the wastewater before it reaches the watercourses. This can be done by infiltration directly into the soil present at the site or by constructing sand-filter trenches. To prevent eutrophication it is necessary that the soil of these systems has a good capacity for holding phosphorus. To be sure of this it is necessary to know which soil factors affect the phosphorus sorption.

A. S. Eikum and R. W. Seabloom (eds.), Alternative Wastewater Treatment, 145–152.

SOIL FACTORS AFFECTING PHOSPHORUS SORPTION

It is generally accepted that most of the phosphorus sorption capacity in soil is due to the soil's finest fractions. This is to be expected, since the active surface area increases with decreasing particle size. Because of this a positive correlation is often found between phosphate sorption and clay content. This correlation can also be due to iron and aluminum on the surface of the clay minerals rather than the pure minerals. The importance of organic matter content on the phosphate sorption is ambiguous. Organic matter can act on phosphate sorption in two ways, either by sorbing phosphate or by blocking sorption sites on inorganic particles.

It is difficult to separate the effects of single parameters such as parent material, texture and organic matter content, since these often are intercorrelated and also correlated with other parameters active in phosphate sorption, e.g. iron and aluminum.

About 100 years ago it was shown that artificial iron and aluminum hydroxides could sorb considerable amounts of phosphate from a solution. That iron and aluminum compounds are very important for the sorption of phosphate in soils has been clearly reported in a great number of publications. This fact can be shown by four types of experiments:

a) The phosphate sorption increases with increasing amounts of Fe- and Al-compounds in the soil,
b) there is increased recovery of Fe- and Al-phosphates after addition of soluble phosphates to the soil,
c) there is increased sorption of phosphates after addition of Fe- and Al-compounds to the soil, and
d) there is decreased sorption of phosphate after removing or blocking the Fe- and Al-compounds.

Different extractants are used to find the contribution to the sorption from the distinct iron and aluminum compounds. Because the extractants are not quite selective the results will differ with the soil type. This is expected because the amorphous Fe- and Al-compounds are the most active in the sorption. Some experiments indicate that aluminum is most active in the first phase of the sorption, but that the sorbed phosphate will turn to iron phosphates over time. However, it is difficult to put forward a general statement about the relative contribution to the sorption from iron and aluminum compounds. Additional experiments to illustrate the importance of adding different Fe- and Al-compounds to soil wastewater treatment systems will be of interest.

A negative correlation between phosphate sorption and pH is found in soils. It is, however, difficult to find soils where only the pH differs and therefore in most of the experiments the relationship between phosphate sorption and pH is difficult to establish. Experiments with gibbsite and goethite have given sorption isotherms with breaks (maximum) at the acids' pKa-values. Even if the surface is negatively charged, undissociated acid molecules can be sorbed as long as they can dissociate on the surface and give protons. This process takes place most easily near the acids' pKa-values. Also other experiments with pure oxides and clay minerals have shown well-defined sorption maxima at given pH values.

Generally, phosphate sorption seems to decrease with decreasing redox potential. This is explained by the change of Fe^{3+} to Fe^{2+}. Soils earlier exposed to reducing conditions have shown great capacity to sorb phosphate by formation of new, amorphous iron(III)compounds. It is important to realize that reducing conditions can release earlier sorbed phosphate.

Addition of different organic anions clearly influences phosphate sorption. The organic anions can block the phosphate sorption sites to differing degrees. How effective the anions are, depends upon the relative stabilities of the Fe(Al)-organic anion complexes and the Fe(Al)-phosphate complexes. The complex stability is pH-dependent so the different organic anions are effective at unequal pH values. It has also been shown that organic anions can prevent crystallization of Al-hydroxides and therefore increase the phosphate sorption compared with systems without organic anions. Also wastewater contains organic compounds and it should be of interest to know how these influence the phosphate sorption in soil wastewater treatment systems. A few new experiments should be only needed in addition to those reported in the literature.

Phosphate sorption is fast in the beginning and slows down to reach a pseudo-equilibrium after some hours or days, but the sorption continues over very long time. The solubility of sorbed phosphate decreases over time.

The effect of temperature on phosphate sorption depends on the system and on which part of the reaction being studied. But within reasonable temperature ranges the effect seems not to be great. Since the effect of temperature on phosphate sorption varies from system to system, this should be more closely examined in soil wastewater treatment systems. In such systems the annual temperature variation can be considerable.

It is shown in batch experiments that the solution:soil ratio and the shaking intensity are of importance for the phosphate sorption. Too strong shaking can break down aggregates and expose new surfaces for sorption. The higher the phosphate: soil ratio is, the greater the phosphate sorption per unit weight of of soil. Increased ionic strength has increased phosphate sorption, but in addition, the effect of cation species must be considered.

Some scientists have claimed that drying and wetting of the soil has led to complete or partial recovery of the soil's sorption capacity, but this is not generally accepted.

PHOSPHATE SORPTION ISOTHERMS AND SORPTION MECHANISMS

Phosphate sorption in soil has been fitted to several sorption equations. Among these the Langmuir and Freundlich equations are undoubtedly the most widely used. The Langmuir equation has the advantage that a sorption maximum can be calculated. On the other hand this equation presupposes the sorption energy not to vary with the degree of surface saturation.

The experimental data generally fit relatively well with the sorption equations for narrow concentration ranges. To cover wider concentration ranges a Langmuir equation has been used for the different parts of the sorption isotherm. Efforts have been made to improve the sorption equations by expanding them with new variables and constants. Some investigations have tried to explain the nature of the sorption mechanisms by interpretation of the sorption equations, but others have been very critical of these attempts. In addition to the sorption equations, single point indexes have been proposed to determine the phosphate sorption capacity of the soil.

Concerning the sorption mechanisms of phosphate in soils, the literature can be roughly divided into two groups. One group claims that the phosphate sorbs in three different regions of the isotherm. The two first regions include chemisorption, while sorption in the third region is of more physical potential-determined type. It is assumed that $-OH_2$ is exchanged with phosphate in region I and with -OH in region II. Much of this partition is based on interpretation of Langmuir's isotherm. The other group claims that sorption is due to only one mechanism. They are of the opinion that the sorption affinity decreases as the sorption increases. This effect is obvious because specific sorption of anions increases the negative charge on the soil particles. The last group also is of the opinion that it is impossible to distinguish between chemisorption and other types of sorption because of gradual changes.

The phosphate kinetics have been described by different kinetic models. It is necessary to have adequate information about the soil to make use of such models. It is difficult to recommend any specific model because they have been tried out for different systems. The models are usually tested on phosphate sorption in columns with homogeneously packed soil. A few have been further tested in full scale wastewater treatment systems. In order to use soil as a renovation system for phosphorus in wastewater we need information not only about how much phosphorus the soil can sorb, but also how fast the sorption reaction takes place. Some of the kinetic models have been shown to be useful in this connection and can be a helpful aid in planning soil wastewater treatment systems.

INVESTIGATIONS STRESSING PHOSPHORUS RENOVATION OF WASTEWATER

Sorption of phosphate from wastewater has been studied both in soil column experiments and in soil wastewater treatment systems to measure the phosphate sorption under more practical conditions. The column experiments have shown that the phosphate sorption capacity is dependent on the soil factors mentioned earlier.
A weak point with all short-term experiments is that they do not include long-term sorption. Nevertheless, several column experiments have been running long enough so that the soil's sorption capacity has been found to exceed the capacity calculated from sorption isotherms. The soil column experiments have clearly shown that use of unsuited soils gives low phosphate sorption. Addition of compounds containing iron and aluminum has improved the phosphate sorption considerably.

Also many investigations with addition of wastewater to different soil types under field conditions have been carried out. They show the highest phosphate sorption in the soil horizons rich in iron and aluminum, e.g. the B-horizons. In the C-horizon and other unsuited soils the phosphate sorption capacity is low. It is therefore beneficial to infiltrate the wastewater as high up in the profile as possible or use soil from well-developed B-horizons in construction of sand-filter trenches. Also the sorption capacity under field conditions has been found higher than that calculated from sorption isotherms.

CONCLUSION

It has clearly been shown in the literature which soil factors influence the phosphorus sorption in soil. A lot is known about the sorption mechanisms and it has also been shown how the soil's sorption capacity can be determined. Most of these methods are too laborious for selection of soil to be used in wastewater treatment systems. It is very important to find a simple and accurate sorption index which is related to the soil's real sorption capacity for phosphorus from wastewater.

Another important goal in connection with phosphorus removal from wastewater is to determine the system's operation time. The operation time will be the time before the output concentration from the treatment system reaches a certain fixed value. In the literature this problem is handled from different points of view, but studies of the sorption kinetics seem to be important in this connection. To solve this problem there is a need for more investigations comparing laboratory sorption studies and measurements in the field.

REFERENCES

Stuanes, A.O.: 1981, "Fosforbinding i jord. En litteraturstudie til støtte for undersøkelser av fosforfjerning fra avløpsvann i jordrenseanlegg." (Phosphorus sorption in soil. A review to support investigations of phosphorus cleaning of wastewater in soil wastewater treatment systems.) Norwegian Forest Research Institute. Division of Forest Ecology. 112 pp., 198 references.

MODELING PHOSPHORUS SORPTION AND MOVEMENT IN SOILS IN RELATION TO SEPTIC TANK LEACH FIELDS

Carl G. Enfield, Ph.D.

Soil Scientist, U.S. Environmental Protection Agency,
Robert S. Kerr Environmental Research Laboratory,
Ada, Oklahoma 74820, USA

Eutrophication of lakes and streams is a natural process which can be accelerated by the movement of nutrient rich wastewater into waterways. Phosphorus (P) has been identified as the nutrient most likely limiting primary productivity in lakes and streams. A prerequisite to understanding the impact of land application wastewater treatment is an assessment of the interaction of phosphorus and soil constituents. The nature of phosphrus soil reactions is complex, as evidenced by the large number of papers in the literature (1) (2).

The soil is a dynamic chemical and biological system in a state of constant flux and disequilibrium. Interpretation of laboratory results to field situations should be made with caution. This paper uses existing data to describe the chemistry and physics of phosphorus in a soil water system emphasizing wastewater application to soils. The paper begins with a description of the chemical forms of P and associated soil constituents and then the kinetics of the reaction. After describing the soil P interaction, P transport is compared to measured P activity in dynamic systems.

A. S. Eikum and R. W. Seabloom (eds.), Alternative Wastewater Treatment, 153–168.

Phosphorus in raw municipal wastewater is generally divided into two broad categories, inorganic P and organic P. The organic P, which typically constitutes 50 percent of the total applied phosphorus, is generally filtered near the point of application. It is considered in this discussion only after mineralization (conversion from organic form to inorganic form).

CHEMICAL STABILITY MODELS

Most inorganic phosphates found in the soil regardless of its origin can be classified into three groups: (1) those containing calcium phosphates; (2) those containing iron and aluminum phosphates; and (3) those combining with silicate materials. The relative importance of these compounds can be roughly correlated to the pH of the soil environment. In acid soils, iron and aluminum phosphates are more common. In basic soils, calcium phosphates predominate. Figure 1 gives a qualitative description of the fate of phosphates in soil versus pH (3).

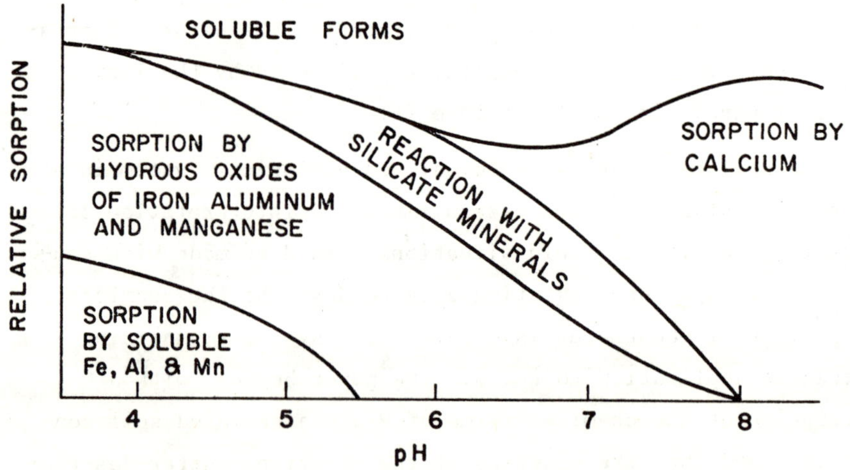

Figure 1. Relative fate of phosphorus in mineral soils as a function of pH.

By studying the solubility products of some of the more important iron, aluminum, and calcium compounds commonly found in soil, it is possible to quantitatively project stability diagrams for the concentration of phosphorus in soil solution.

The equilibrium isotherms, plotted in Figure 2 for selected phosphate compounds, were developed from Gibbs free energies in Table 1. The values selected for the Gibbs free energies were from a single source (4) so that they would be consistent with each other. There is a wide range in the values, and the projected concentrations could range several orders of magnitude if the reported extremes were used. Several assumptions were made in calculating Figure 2: (1) calcium was assumed to be 0.01 molar; (2) the ionic strength was assumed to be 0.026 mole liter^{-1} (ionic strength equals 0.013 times the electrical conductivity in millimhos cm^{-1}); (3) variscite ($AlPO_4 \cdot 2H_2O$) was assumed to be in equilibrium with amorphous aluminum hydroxide ($Al(OH)_3$); (4) strengite ($FePO_4 \cdot 2H_2O$) was assumed to be in equilibrium with iron hydroxide ($Fe(OH)_3$); and (5) in the pH range of interest (HPO_4^{2-}) and ($H_2PO_4^-$) were the only phosphate species significantly affecting the P concentration. When possible, it would be better to know the activity of the iron, aluminum, calcium, and oxidation state rather than estimate activity from an assumed associated compound.

SORPTION-DESORPTION MODELS

There is confusion as to the terminology used in the literature describing the loss of phosphorus from solution. Some, (5) (6) (7) (8) (9), choose to describe their data using solubility product theory, as above, assuming a precipitation type of process. Other researchers (10) (11) (12) attempt to describe the process using sorption isotherms. The most commonly used sorption expression for phosphorus is the Langmuir equation (13) which can be written in its simplest form as

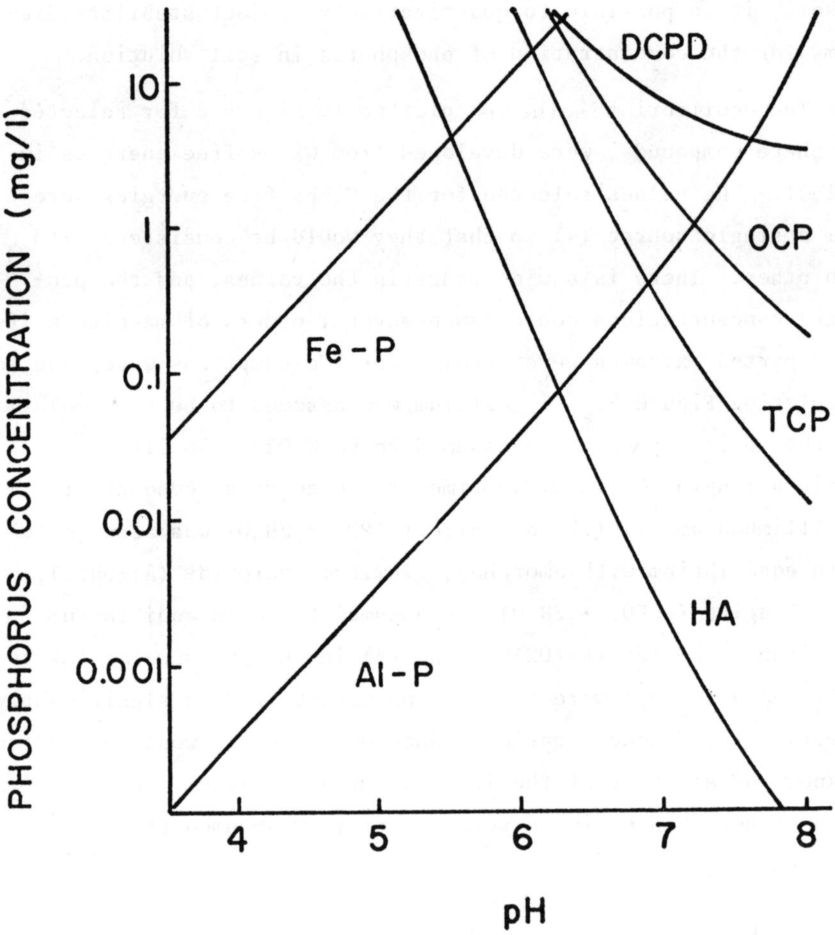

Figure 2. Stability diagram for selected phosphate compounds. Fe-P is the stability line for strengite ($FePO_4 \cdot 2H_2O$) in equlibrium with $Fe(OH)_3$. Al-P is the stability line for variscite ($AlPO_4 \cdot 2H_2O$) in equilibrium with amorphous $Al(OH)_3$. The calcium phosphates dicalcium phosphate dyhydrate (DCPD) $Ca(H_2PO_4)_2 \cdot H_2O$, octacalcium phosphate (OCP) Ca_8H_2 $(PO_4)_6 \cdot 5H_2O$, tricalcium phosphate (TCP) $Ca_3(PO_4)_2$, and hydroxyapatite (HA) $Ca_{10}(OH)_2$ $(PO_4)_6$ are calculated for 0.01M Ca and an ionic strength of 0.026 mole $liter^{-1}$.

TABLE 1. REPORTED GIBBS FREE ENERGY OF SELECTED COMPOUNDS*

	ΔG_F^o
Al^{3+}	-117.33
$AlOH^{2+}$	-167.17
$Al(OH)_2^+$	-218.02
$Al(OH)_3^o$	-266.94
$Al(OH)_3$ (amorphous)	-274.21
α - $Al(OH)_3$ (bayerite)	-275.78
γ - $Al(OH)_3$ (gibbsite)	-276.43
$Al(OH)_3$ (nordstrandite)	-276.3
$AlPO_4$ (berlinite)	-388.50
$AlPO_4 \cdot 2H_2O$ (variscite)	-505.97
Ca^{2+}	-132.52
$CaCO_3^o$	-263.00
$CaCO_3$ (calcite)	-270.18
$CaPO_4^-$	-386.51
$CaHPO_4^o$	-398.29
$CaH_2PO_4^+$	-406.28
$CaHPO_4$ (monetite)	-403.96
$CaHPO_4 \cdot 2H_2O$ (brushite)	-516.89
$Ca(H_2PO_4)_2 \cdot H_2O$ (c)	-734.48
α - $Ca_3(PO_4)_2$ (c)	-922.70
β - $Ca_3(PO_4)_2$ (tricalcium phosphate)	-927.37
$Ca_8H_2(PO_4)_6 \cdot 5H_2O$ (octacalcium phosphate)	-2942.62
$Ca_{10}(OH)_2(PO_4)_6$ (hydroxyapatite)	-3030.24
$FeOH^+$	-69.29
$FeOH^{2+}$	-57.72

(continued)

TABLE 1. (continued)

	ΔG_F^o
$Fe(OH)_2$ (c)	-117.58
$Fe(OH)_3$ (amorphous)	-169.25
$Fe(OH)_3$ (soil)	-170.40
α - FeOOH (geothite)	-117.42
$FePO_4 \cdot 2H_2O$ (strengite)	-398.59
$Fe_3(PO_4)_2 \cdot 8H_2O$ (vivianite)	-1058.36
PO_4^{3-}	-245.18
HPO_4^{2-}	-262.03
$H_2PO_4^-$	-271.85
$H_3PO_4^o$	-274.78
H_2O (l)	-56.69

* From Sadiq and Lindsay (4)

$$\frac{C}{S} = \frac{1}{KS_{max}} + \frac{C}{S_{max}} \qquad (I)$$

where:

C = solution concentration of phosphorus (mg/l)

S = adsorbed phosphate of concentration C (μg P/g soil)

S_{max} = adsorption maximum which is equivalent to a monolayer of phosphate on the adsorbing surface (μg P/g soil)

K = a constant related to the bonding energy of the chemical.

The second most common equation is the Freundlich equation

$$S = K\,C^{1/n}. \qquad (II)$$

where:

S = phosphate adsorbed (μg P/g soil)

K = constant (not necessarily equal to Langmuir K)

C = equilibrium concentration of phosphorus (mg P/l)

$1/n$ = constant.

The Freundlich equation can be shown to be equivalent to a multiple layer Langmuir equation. Both of the methods have successfully described phosphorus behavior in soil. Both of these equations, as generally written, assume phosphorus reactions are instantaneous and at equilibrium with the soil. If these equations alone were used to predict the movement of phosphorus in soil, one would expect phosphorus movement to appear as if it were in a chromatographic column. The chemical, when applied, would move through the soil as a slug, having roughly a bell shaped distribution to account for dispersion. This type of description does not appear consistent with most reported breakthrough data in the literature.

KINETIC MODELS

The disappearance of P from soil solution occurs as a fast initial reaction followed by a much slower reaction. Investigators believe sorption is the primary process during the initial rapid reaction, and precipitation to relatively insoluble phosphates controls the slow reaction (14). This type of observation can be seen in Figure 3 where a series of phosphorus additions to one soil sample were monitored with time. The quick response takes place in the first few hours followed by a gradual change for a long period of time. Considering the time frame for most environmental problems, this rapid sorption can be approximated as an instantaneous reaction. Following the initial sorption, researchers (14) (15) (16) (17) (18) and (19) have tried to describe the response using first order kinetics

$$\frac{\partial S}{\partial t} = K_p C \qquad \text{(III)}$$

where K_p is the rate coefficient. One problem with this approach is that different kinetic rate coefficients are obtained when

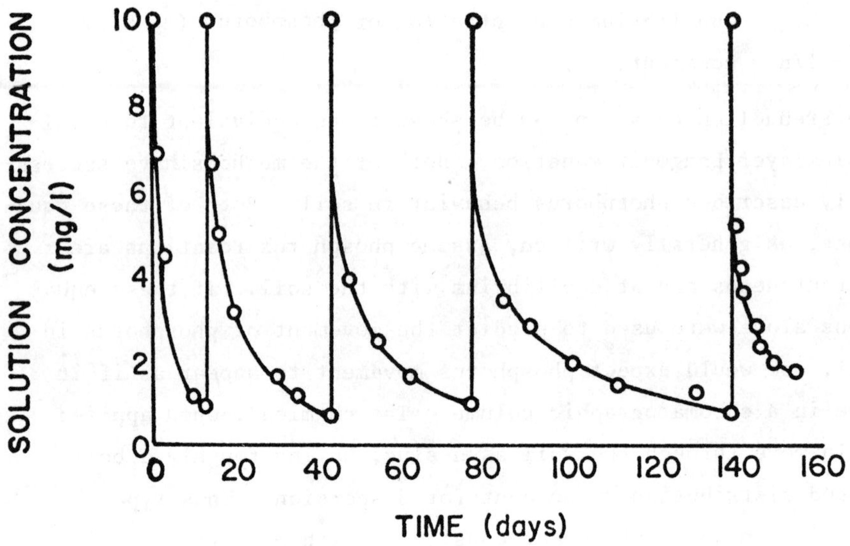

Figure 3. Response to a series of phosphate additions to a single soil sample.

different initial conditions are used for the same soil sample (Figure 4). As written in Equation III, the driving force for the reaction is the concentration. Perhaps a better way to describe the equation would be

$$\frac{\partial S}{\partial t} = \sum_{j=1}^{n} K_j \,(C - CE_j) \qquad \text{(IV)}$$

where: CE_j is the equilibrium concentration such as observed in Figure 1, and j is an index to the many different compounds which potentially could form.

In Equation IV the driving force would be the difference between the concentration of phosphorus in solution and the concentration that would exist if the system were in equilibrium with a given phosphate compound. Equation IV would permit multiple compounds forming simultaneously when the solution concentration is high. As the solution concentration declines, the

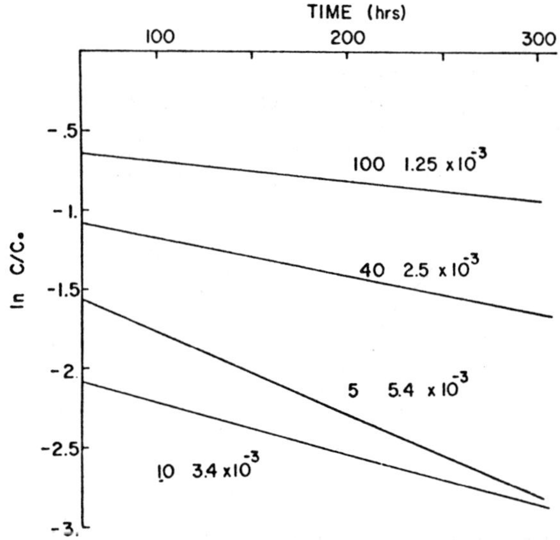

Figure 4. For a soil pH 5.5 a first order rate plot for initial concentrations of 5, 10, 40, and 100 mg/l the specific rate coefficient is referenced to the solution rather than the solid phase. The solid phase rate coefficient would be ten times the solution phase rate coefficient for this particular experimental data set.

more soluble compounds would dissolve leaving only the less soluble compounds. Observations of multiple compounds forming simultaneously have been reported in the literature (20) and support the above description.

Researchers (19) (21) (22) observed the rate coefficient was pH dependent. Their observations (Figure 5) show the slowest rate occurs near neutral conditions and range from 10^{-1} at pH 3.5 and 8.5 to 10^{-4} hrs^{-1} at pH 7. The range covers most observations for either batch or column studies reported in the literature.

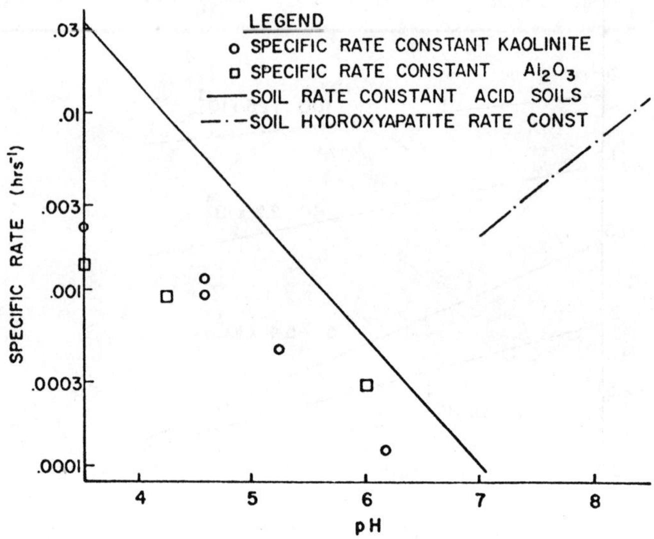

Figure 5. Specific rate as a function of pH. Note: Data for kaolinite and Al_2O_3 (19) is referred to the liquid phase; data on acid soils (21) and hydroxy-apatite (22) is referred to the solid phase.

TRANSPORT MODELS

Convective-diffusive transport of phosphorus through soil is described assuming local equilibrium for a linear sorption process and first order kinetics to describe the chemical precipitation. In one dimensional differential form, the equations for the liquid phase follows

$$\frac{\partial\,(\Theta\rho_L\, C_L)}{\partial t} = D\,\frac{\partial^2(\Theta\rho_L\, C_L)}{\partial x^2} - V\,\frac{\partial\,(\Theta\rho_L\, C_L)}{\partial x} - K_L\,\Theta\rho_L\, C_L$$

$$+ K_s\,\rho_s\, C_L - K_{pL}\,\Theta\rho_L\, C_L \qquad \text{(V)}$$

where:

C_L = concentration (μg/g) in the liquid phase

D = apparent hydrodynamic dispersion coefficient (cm^2/hr)

K_{pL} = rate of chemical precipitation in liquid phase (hrs^{-1})

K_L = rate of sorption from the liquid phase to the solid phase (hrs^{-1})

K_s = rate of desorption from the solid phase to the liquid phase

V = interstitial liquid velocity (cm/hr)

x = distance along flow path (cm)

Θ = volumetric water content ratio (cm^3/cm^3)

ρ_L = density of the liquid (g/cm^3)

ρ_s = density of the solid phase (g/cm^3).

In the solid phase

$$\frac{\partial (\rho_s C_s)}{\partial t} = K_L \Theta \rho_L C_L - K_s \rho_s C_s - K_{ps} \rho_s C_s \quad \text{(VI)}$$

where:

C_s = concentration in the solid phase (μg/g)

K_{ps} = rate of precipitation in the solid phase.

If we assume:

(1) Θ, ρ_L, ρ_s, P, and temperature are constant during the experiment,

(2) the sorption-desorption process remains in local equilibrium,

(3) the precipitation process takes place only when it has previously been sorbed on the solid phase, and

(4) the soil column behaves as if it were infinitely long.

Then from assumption 2

$$C_s = \frac{K_L \Theta \rho_L}{K_s \rho_s} C_L \quad \text{(VII)}$$

by defining the soil-liquid partition coefficient K_d as $K_L \Theta \rho_L / K_s \rho_s$

$$C_s = K_d C_L \quad \text{(VIII)}$$

The total solution can be obtained by summing equations V and VI giving

$$\frac{\partial C_L}{\partial t} = D \frac{\partial^2 C_L}{\partial x^2} - V \frac{\partial C_L}{\partial x} - \frac{\rho_s}{\Theta \rho_L} \left(\frac{\partial C_s}{\partial t} + K_{ps} C_s + \frac{K_{pL} C_L}{\rho_s} \right) \quad \text{(IX)}$$

by letting $\frac{\partial s}{\partial t} = K_{ps} C_s + \frac{K_{pL} \Theta C_L}{\rho_s} + \frac{\partial C_s}{\partial t}$ and assuming the density $(\rho_L) = 1.0$, then

$$\frac{\partial C_L}{\partial t} = D \frac{\partial^2 C_L}{\partial x^2} - V \frac{\partial C_L}{\partial x} - \frac{\rho_s}{\Theta} \frac{\partial s}{\partial t} \quad \text{(X)}$$

This equation has been solved for several sets of boundary conditions (e.g. (17) (18) (23) (24) (25). One solution to Equation X where K_{pL} is assumed equal to zero is

$$\frac{C}{C_T} \Big|_x = \frac{1}{2} \left(1 - \operatorname{erf} \frac{x - (Vt/R)}{2 \sqrt{D/R}} \right) \quad \text{(XI)}$$

where:

erf = error function

$$C_T = C_o \exp\left(- \frac{\rho_s}{\Theta} t K_{ps}\right) \quad \text{(XII)}$$

C_o = initial concentration (mg/l)

$$R = \text{retardation factor} = 1 + \frac{\rho_s}{\Theta} K_d \quad \text{(XIII)}$$

t = time (hrs)

This equation was shown to fit P breakthrough data (24) as shown in Figure 6. The figure is a projection of breakthrough curves for three columns having lengths of 1, 3, and 3.9 cm, for a soil with a bulk density (ρ_s) of 1.4 g/cc, volumetric water content (Θ) of 0.18, an interstitial velocity (V) of 1.0 cm/hr, a rate of precipitation in solid phase (K_{ps}) of 0.02 hr^{-1}, a linear partition coefficient (K_d) of 4.3, and dispersion coefficient (D) of 0.1 cm^2/hr.

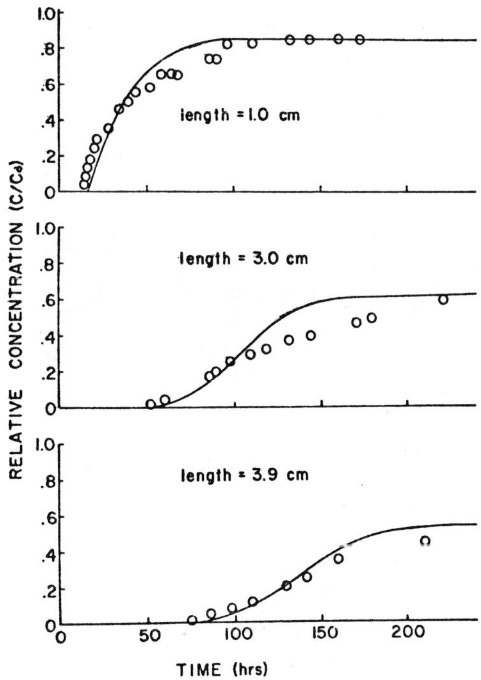

Figure 6. Measured and projected breakthrough curves for phosphorus. The soil bulk density is 1.4 g/cm^3, volumetric water content 0.18, interstitial velocity 1.0 cm/hr, precipitation rate 0.02 hr^{-1}, linear partition coefficient 4.3, and dispersion coefficient 0.1 cm^2/hr.

When designing a leach field for a septic system a major interest is the concentration of phosphorus in solution at a quasi-steady state system, rather than the transient response to a pulse of phosphorus. Equation XII has been used in laboratory studies (17), and Leach et al. (26) used Equation XII to evaluate existing rapid infiltration wastewater treatment systems. Leach et al. assumes $\rho_s K_{ps}/\Theta \geq 0.002$ hr^{-1}, which would permit making conservative estimates of the concentration at any point along the flow path when the initial concentration and interstitial velocity are known (Table 2). The value for the constant selected appears to be reasonably consistent with range of rate coefficients presented earlier.

TABLE 2. COMPARISON OF MEASURED AND PREDICTED PHOSPHORUS CONCENTRATION IN SOIL SOLUTION UNDER RAPID INFILTRATION WASTEWATER TREATMENT SYSTEMS ADAPTED FROM LEACH et al. (26)

Location	Sampling depth m	pH	P Applied mg/l	P Predicted mg/l	P Observed mg/l
Hollister, CA	7.7	7.7	12.4	10	8
Milton, WI	8	7.5	8.2	7	5.9
Lake George, N.Y.	22	6.5 - 7.4	4	1.3	0.3 - 1.1
Vineland, N.J.	1.5	6.6 - 6.9	4.5 - 7.2	4.3	4.3

CONCLUSIONS

The actual mechanisms for phosphorus disappearance from solution is yet to be fully understood. Methods to accurately estimate the ultimate capacity of a soil to immobilize phosphorus are not yet available. With these limitations being recognized, methods are available and have received preliminary testing which project the concentration of phosphorus in solution in a dynamic soil-water system. The results of these predictions are reasonably accurate. Further studies are needed to assess the effect of temperature, pH, and retardation coefficient on the projections. Methods beyond simple sorption isotherms, which underestimate the soil's capacity, are needed to avoid over-restricting the use of the soil as a waste receptor.

REFERENCES

1. Chien, S.H. & W.R. Clayton: 1980, "Application of Elovich equation to the kinetics of phosphate release and sorption in soils." Soil Sci. Soc. Am. J. 44, pp 265-268.

2. Berkheiser, V.E., J.J. Street, P.S.C. Rao & T.L. Yuan: 1980, "Partitioning of inorganic orthophosphate in soil water systems." C.R.C. Critical Reviews in Environmental Control 10, pp. 174-224.

3. Buckman, Harry O. & Nyle C. Brady: 1969, "The Nature and Properties of Soils." 7th Edition. The Macmillan Company. London. 480.

4. Sadiq, Muhammad & W.L. Lindsay: 1979, "Selection of standard free energies of formation for use in soil chemistry." Colorado State University, Technical Bulletin 134. pp. 1069.

5. van Riemsdijk, W.H., F.A. Westrate & H.G. Bolt: 1975, "Evidence for a new aluminum phosphate phase from reaction rate of phosphate with aluminum hydroxide." Nature 257, pp. 473-474.

6. Bennett, A.C. & Fred Adams: 1976, "Solubility and solubility product of dicalcium phosphate dihydrate in aqueous solutions and soil solutions." Soil Sci. Soc. Am. J. 40, pp. 39-42.

7. Subbarao, Y.V. & Roscoe Ellis, jr.: 1977, "Determination of kinetics of phosphorus mineralization in soils under oxidizing conditions." Robert S. Kerr Environmental Research Laboratory, Ada, Oklahoma, EPA-600/2-77-180.

8. Lindsay, Willard L.: 1979, "Chemical Equilibria in Soils." John Wiley & Sons, New York, pp. 162-209.

9. Hsu, Pa Ho: 1979, "Effect of phosphate and silicate on the crystallization of gibbsite from OH-Al solution." Soil Science 127, pp. 219-226.

10. Holford, I.C.R., R.W.M. Wedderburn & G.E.G. Mattingly: 1974, "A Langmuir two-surface equation as a model for phosphate adsorption by soil." J.Soil Sci. 25, pp. 242-256.

11. Shayan, A. & B.G. Davey: 1978, "A universal dimensionless phosphate adsorption isotherm for soil." Soil Sci. Soc. Am. J. 42, pp. 878-882.

12. Sibbesen, E.: 1981, "Some new equations to describe phosphate sorption by soils." J. Soil Sci. 32, pp. 67-74.

13. Langmuir, I.: 1918, "The adsorption of gases on plane surfaces of glass, mica, and platinum." J. Am. Chem. Society 40, pp. 1361-1403.

14. Sawhney, B.L.: 1977, "Predicting phosphate movement through soil columns." J. Environmental Quality 6, pp. 86-89.

15. Cho, C.M., J. Strong & G.J. Racz: 1970, "Convective transport of orthophosphate (P-31 and P-32) in several Manitoba soils." Can. J. Soil Sci. 50, pp. 303-315.

16. Mansell, R.S., H.M. Selim, P. Kanchanasat, J.M. Davidson & J.G.A. Fiskell: 1977, "Experimental and simulated transport of phosphorus through sandy soils." Water Resources Res. 13, pp. 189-194.

17. Overman, Allen R., Ro-Lan Chu & Yung Le: 1978, "Kinetic coefficients for phosphorus transport in packed-bed reactor." J. WPCF 50, pp. 1905-1910.

18. de Camargo, O.A., J.W. Biggar & D.R. Nielsen: 1979, "Transport of inorganic phosphorus in an alfisol." Soil Sci. Soc. Am. J. 43, pp. 884-890.

19. Chen, Y.S.R., J.N. Butler & W. Stumm: 1973, "Kinetic study of phosphate reaction with aluminum oxide and kaolinite." Environ. Sci. and Tech. 7, pp. 327-332.

20. Freeman, J.S. & D.L. Rowell: 1981, "The adsorption and precipitation of phosphate onto calcite." J. Soil Sci. 32, pp. 75-84.

21. Enfield, C.G.: 1978, "Evaluation of phosphorus models for prediction of percolate water quality in land treatment." In: State of Knowledge in Land Treatment of Wastewater, Vol. 1. U.S. Army Corps of Engineers, Cold Regions Research and Engineering Laboratory, Hanover, NH, pp. 153-162.

22. Enfield, C.G., T. Phan, D.M. Walters & R. Ellis, Jr.: "Kinetic model for phosphate transport and transformation in calcareous soils, I. Kinetics of transformation." Soil Sci Soc. Am. J. In press.

23. Enfield, C.G., T. Phan & D.M. Walters: "Kinetic model for phosphate transport and transformation in calcareous soils, II. Lab and field transport." Soil Sci. Soc. Am. J. In press.

24. Enfield, C.G. & R.F. Carsel: 1980, "Mathematical prediction of toxicant transport through soil. Test Protocols for Environmental Fate and Movement of Toxicants." AOAC Washington, D.C. pp. 233-250.

25. Novak, L.T.. D.C. Adriano, G.A. Coulman & D.B. Shah: 1975, "Phosphorus movement in soils: theoretical aspects." J. Environmental Quality 4, pp. 93-99.

26. Leach, L.E., C.G. Enfield & C.C. Harlin, Jr.: 1980, "Summary of long-term rapid infiltration systems studies." Robert S. Kerr Environmental Research Laboratory, EPA-600/2-80-165.

REDUCTION OF ENTERIC MICROORGANISMS IN SOIL INFILTRATION SYSTEMS
Field and Laboratory Experiments

Thor Axel Stenström, Microbiologist
Sven Hoffner, Microbiologist

The National Bacteriological Laboratory, Department of Water Microbiology, S-105 21 Stockholm, Sweden

SUMMARY

The reduction of fecal microorganisms including many pathogenic viruses and bacteria in a soil infiltration unit will depend on various factors in interaction with each other. The soil characteristics, composition of the infiltrated water and characteristics of the microorganism itself are all of great importance. This paper deals with some of the factors of importance for microbial survival and transportation in a soil infiltration unit. The purpose is to increase the knowledge and understanding of these factors and their complex interactions. This could be of great practical value when constructing on-site wastewater treatment systems.

INTRODUCTION

Knowledge of the reduction mechanisms of fecal microorganisms in soil and groundwater is of great importance when estimating the risk of microbial ground water contamination and waterborne diseases via drinking water wells.

A. S. Eikum and R. W. Seabloom (eds.), Alternative Wastewater Treatment, 169–181.

Soil is extensively used as a natural wastewater treatment system. In Sweden, 20,000-25,000 wastewater infiltration units are built every year. Considering that groundwater is used as drinking water source for about 47 percent in towns and villages and for nearly 100 percent in the countryside, it is very important that these units are constructed and maintained in a safe way. To counteract microbiological contamination of groundwater a minimal retention time of 60 days and a minimal distance of 50 meters are recommended between a wastewater infiltration unit and a drinking water well (1).

Since the number of microorganisms that is required to give an infection (infective dose), differs between microbial species and is dependent on the host (age, nutritional status, immunological defence), the risk varies greatly. The effort must be to keep the groundwater free from pathogenic microorganisms and thereby ensure a safe drinking water source. The two key factors when estimating the concentration of microorganisms in groundwater after a certain time and at a certain distance from a wastewater infiltration unit, are the survival of the mircoorganisms and their capacity to penetrate the soil layer. The penetration will depend on interactions between microorganisms and soil, e.g. if the organisms will be filtered away or adsorbed to the soil particles.

Several factors affect the survival and transportation of microrganisms (Figure 1). Many of these will interact with each other in various ways. This paper will deal with some of these factors, based on experiments made at the laboratory or in the field.

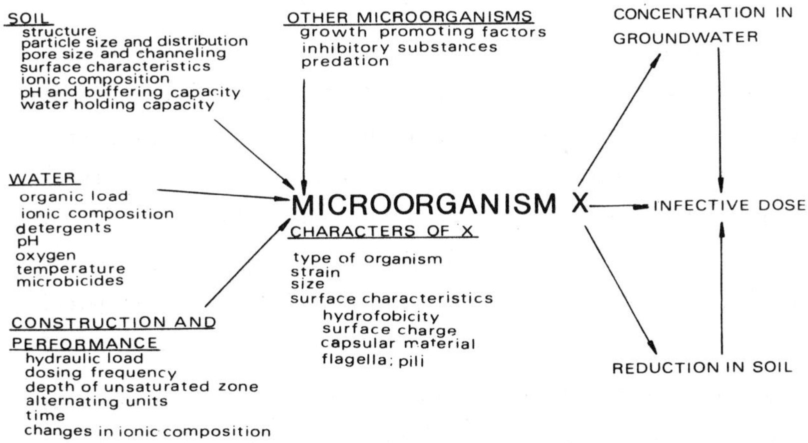

Figure 1. Factors affecting the survival and transportation of enteric microorganisms in soil.

CHARACTERISTICS OF THE MICROORGANISMS

Size

The size is probably of less importance for reduction of bacterial and viral species. A sandy soil with a particle diameter of 0.15 mm will have pores greater than 20 μm (2), and when compared with the size of bacteria (0.5-5 μm) and viruses (about 10-100 times smaller than bacteria), it can be seen that these organisms will not be filtered.

However, parasites and their ova or cysts are bigger and will normally be filtered out effectively from the percolating water. This has been shown in several investigations and recently confirmed in a Norwegian investigation (3).

Hydrophobicity and Surface Charge

Of greater importance in explaining differences between bacterial strains in their ability to be adsorbed to the soil particles are their surface characteristics. The outer membrane of bacteria and wall associated structures like capsules, slime layers, and fimbriae are directly involved in the nonspecific adsorption. These structures expose different charged positive and negative ionic groups which will be involved in an interaction directly to the soil particles as well as indirectly to the attracted macromolecules and counter ions. The exposed surface structures also express hydrophobic/hydrophilic properties which will result in a hydrophobic interaction with soil particles.

An estimation has been made of the surface hydrophobicity and charge of a selected number of bacterial strains, following the method of Kjellberg et al. (4). Some of the results are shown in Figure 2, presented as percentage adsorption.

The negative surface charge is normally high, while a great variability occurs between different strains for their exposed positively charged groups and hydrophobicity.

In our system the three Shigella strains have both a lower surface hydrophobicity and lower positive and negative surface charge than the Escherichia coli strains (fecal indicator), which will lead to a more rapid penetration through a soil. The Shigella strains might therefore in spite of shorter survival be a significant health risk in groundwater.

Streptococcal strains detected after filtration through soil, have a lower hydrophobicity than other strains, which also indicates a more efficient transportation. Research is under progress in the National Bacteriological Laboratory to correlate surface characters with different soil characteristics.

Survival

The survival of fecal microorganisms has been the subject of several reports during the last decades. These studies have recently been reviewed (5). The survival of fecal microorganisms

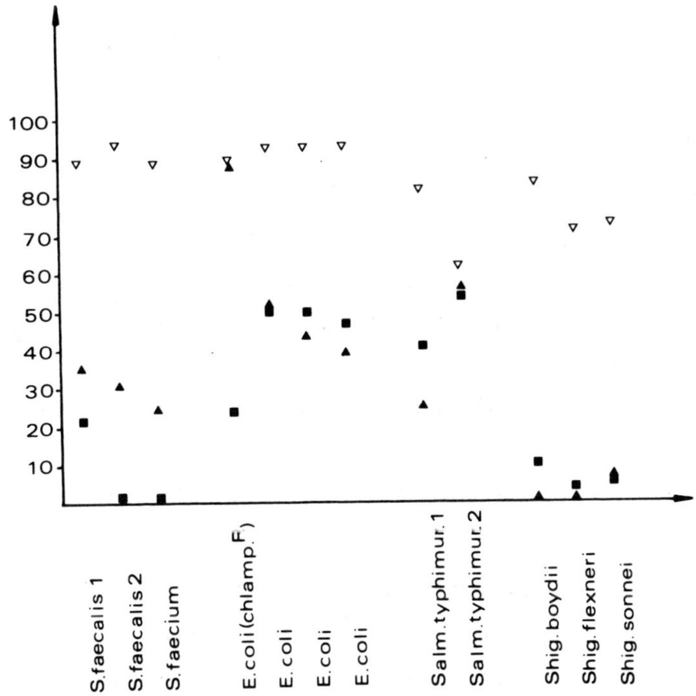

Figure 2. Adsorption of enteric bacteria on octyl-Sepharose (hydrophobic interaction ■) and negatively and positively charged ion exchange resins (estimate the positively ▲ and negatively ▽ charged surface groups respectively). The values are expressed as percentage adsorption on the resins.

differs between species. Temperature, soil type and the effect of its normal flora have been evaluated (6). Several Salmonella and Yersinia strains have a longer survival than E.coli (expressed both as $t_{1/2}$ and total time of detection). Shigella and Campylobacter are normally killed more rapidly. Some Streptococcus strains as well as bacteriophages have a much longer survival than the rest of the test organisms. This illustrates the complexity of a wastewater infiltration unit when looked at from a microbiological view-point. It also pinpoints the necessity to use different indicator organisms in risk assessments. The use of the coliform group only may lead to an underestimation of the possible survival of pathogenic microorganisms.

OTHER MICROORGANISMS

The normal microflora of soil infiltration systems will affect the fecal microorganisms in many ways. By direct predation e.g. from protozoa, or by excreting inhibitory substances they will reduce the number of fecal microorganisms with time. However, the reduction of allochthonous microorganisms will also depend on the physico/chemical micro-environment in the soil infiltration system. The optimal survival under laboratory conditions has therefore been compared with more authentic conditions.

In Figure 3 the survival of inoculated E.coli, Streptococcus faecalis and Salmonella typhimurium is shown in soil samples from different depths in a model sandfilter trench with normal microorganisms and dosed with synthetic wastewater. Further information on the infiltration unit in this study can be found in Andersson et al. (7). The longest survival of E.coli and S.typhimurium could be seen in the top layer of the unit, while a rapid decrease occurred at greater depths. For S.faecalis the opposite was found at greater depths. The same inoculum was used for all levels. The experiments have been repeated in other systems with the same results. In Figure 4 this is exemplified with a model column system with different degrees of pretreatment and with a high hydraulic load. These results are less clear for E.coli, but S.typhimurium exhibits a longer survival in the surface layers, while S.faecalis shows longer survival deep in the column. A general tendency of longer survival was seen in the column called C3, which had the most far-reaching pretreatment (both mechanical, biological and chemical). This treatment will naturally reduce both the numbers of antagonistic

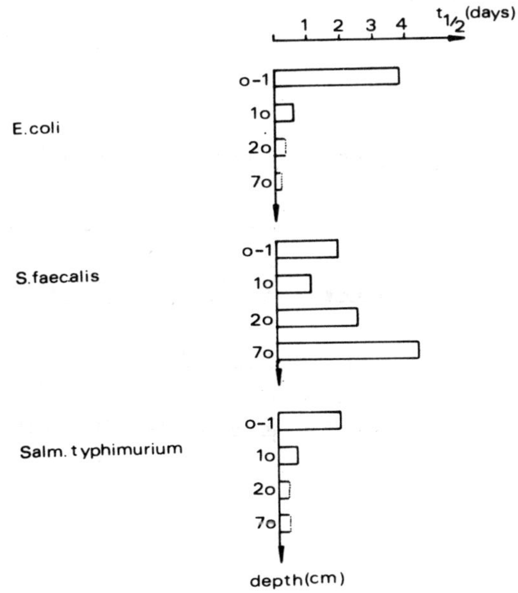

Figure 3. Reduction (expressed as half life, $t_{1/2}$, for test population) of E.coli, S.faecalis and S.typhimurium in soil from different depths in a model sandfilter trench. The same amount of bacteria were inoculated at time zero.

microorganisms and the amount of different compounds. A "predation potential" is presented for S. typhimurium in Figure 5 which correlates with the survival of this organism.

SOIL

The survival and transportation of different bacterial species have been reported (3, 6, 7). Survival of different enteric test organisms are longest in silt or sand. In other types of soils (morains) a faster reduction was seen. This was more pronounced at the higher of the tested temperatures (20 ^{o}C). At 4 ^{o}C a long survival was seen in all soils tested, especially for Streptococcus sp. This is exemplified in Figure 6.

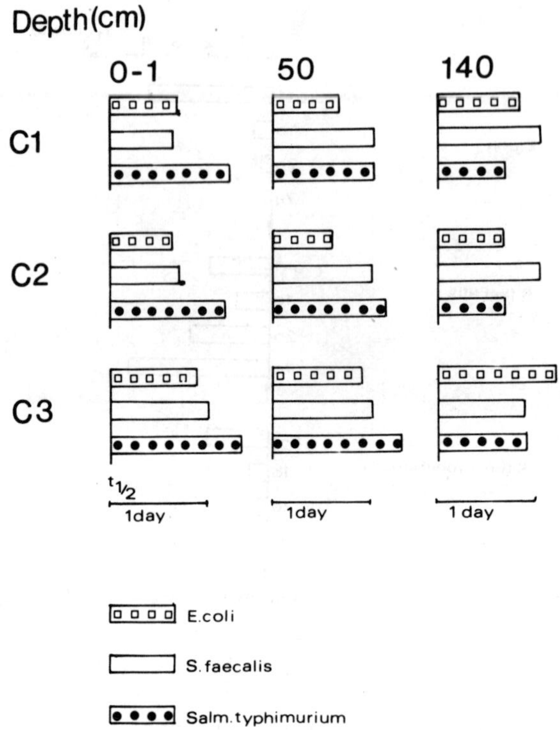

Figure 4. Reduction (expressed as half life, $t_{1/2}$, for test population) of E.coli, S. faecalis and S.typhimurium in soil from different depth in a column infiltration system with different pretreatment. C1 = mechanical pretreatment, C2 = mechanical and biological pretreatment, and C3 = mechanical, biological and chemical pretreatment.

The hygienic risks following a longer survival in silty soils are counteracted by a higher adsorption and thereby reduction of fecal microorganisms. This is partly due to the smaller particle sizes which will expose a larger surface area and facilitate adsorption. Channeling might be a problem of special concern. In most systems investigated so far, tracer organisms that have been added to the inlet water, have been collected at the outlet within a very short time. This constitutes a malfunction of the

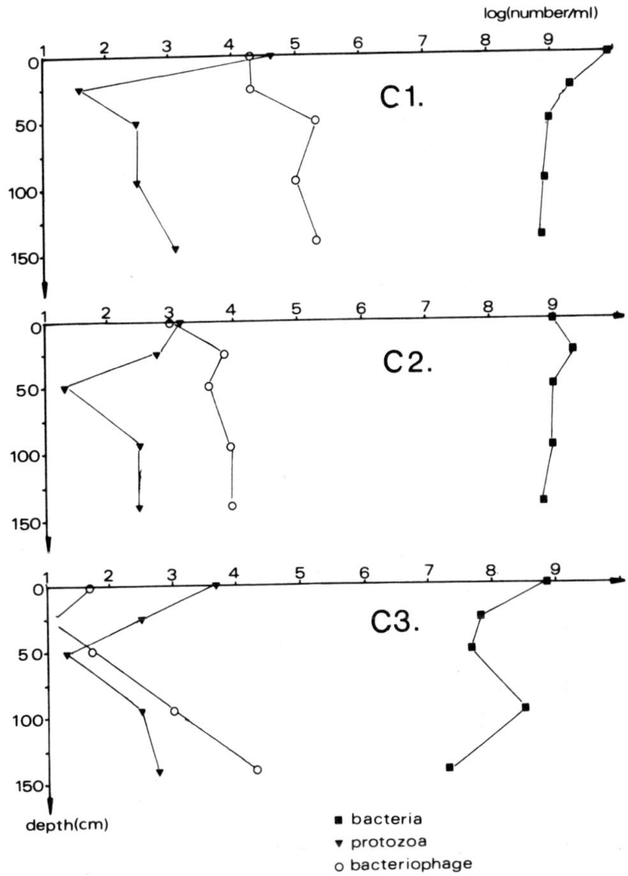

Figure 5. "Predation potential" for S. typhimurium.

■ bacteria, epifluorescent count with ethidiniumbromid.
▼ number of protozoa, able to predate on S.typhimurium, counted by titration with the bacteria as receiving strain.
O bacteriophages able to lyse S.typhimurium. Expressed as plague forming units (pfu).

system e.g. the channeling did not give the penetrating microorganism the close contact with soil particles which was necessary for an effective adsorption.

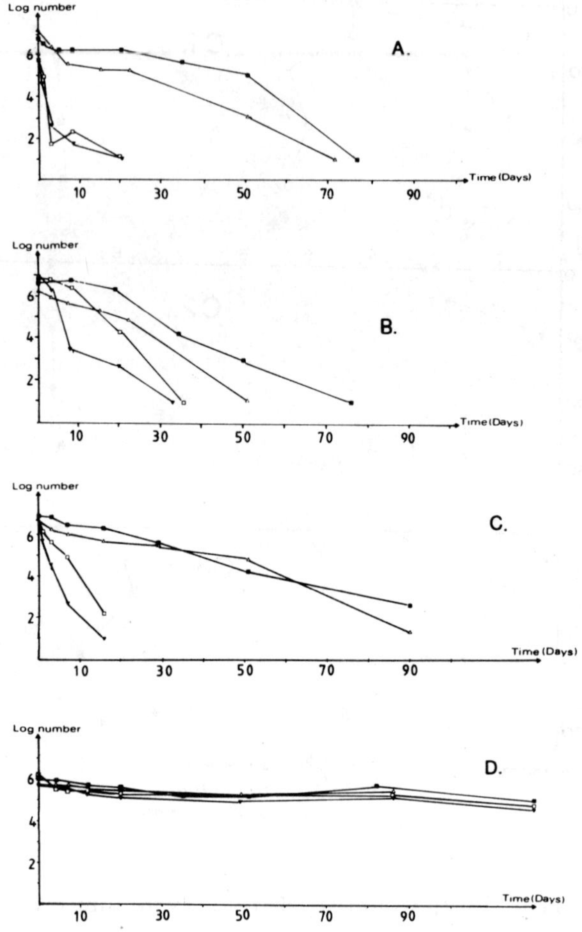

Figure 6. Survival of E.coli (A and C) and S.faecalis (B and D) in silt (■), sand (△) and two morains (□ , ▼) at 20 °C (A and B) and 4 °C (C and D). Field capacity was maintained at 70 percent.

WATER

Lower temperature always gives a longer survival in water or soil. It might under favorable conditions be extended for several months with only a slow reduction. The presence of trivalent positive ions (and some of the divalent ions) will favor the adsorption of bacteria and viruses to the soil particles with a higher reduction as a result. However, this is a reversible process and a high degree of desorption will take place if the ionic concentration is suddenly lowered. This may be the case after heavy rains, or sudden dosing of large volumes of stormwater, which would then give a sudden rise of microorganisms in the groundwater.

The presence of detergents in the water (e.g. gray water) will also favor the penetration of microorganisms through the soil, without drastically altering their survival. However, in the presence of di- and trivalent ions a flocculation of the detergents will occur in different concentration intervals (due to type of detergent) that give the opposite effect: clogging, and a reduction of the flow of water and microorganisms through the soil. The same flocculation process might occur with organic material, polysaccharides as well as the capsular material present at the surface of different bacteria. The flocculation of bacterial polysaccharides might therefore be of importance in reducing the number of allochthonous microorganisms in soil.

CONSTRUCTION AND PERFORMANCE

The hydraulic load is a factor of major importance. Under "normal" situations (a load of 5-10 cm/day) a reduction of microorganisms around 100 times can be seen if the soil layer is between 50 and 80 cm. In the column system mentioned above, (C1-3) with different degree of pretreatment the depth of the sand was about 150 cm and the hydraulic load 2 cm/hour. This gave the same reduction as in a comparable model and field system, with "normal load". (Table 1.) However, a partially higher hydraulic load is often seen in soil infiltration units and sandfilter trenches, due to clogging of parts of the units. This will then lead to a higher penetration of the fecal microorganisms in the unit, or to a channeling effect that gives minimal

Table 1. Cumulative reduction of antibiotic resistant tracer bacteria and a bacteriophage in
1. a model sand filter trench
2. a field sandfilter trench, and of normal indicators in a column system with different degree of pretreatment (C1-3, see Figure 4).

	E. COLI		S. FAECALIS		S. P. FO-PHAGE	
	n	mv	n	mv	n	mv
1	4	98.8	4	94.6	2	96.5
2	2	98.4	2	98.3	2	97.4
	TOT. COLIF.		FECAL COLIF.		FECAL STREPT.	
C1	4	98.5	3	99.2	4	97.2
C2	4	93.1	4	94.5	4	93.9
C3	4	96.5	2	91.0	3	96.1

retention and reduction. This has been discussed by Andersson et al. (7). The dosing frequency will also affect the efficiency of an infiltration unit. This has been studied in a column system (3). The hydraulic load of 5 cm/day was divided into 2.4, 4, 8, and 72 portions/day and the penetration of microorganisms through the columns was studied. The best bacterial retention was seen for 4-8 portions/day. The effect of the number of portions/day was less pronounced for the tested coliphages. When the hydraulic capacity was reduced 80-90 percent, no effect of the different dosage frequency could be seen. When a new soil infiltration unit is started, a higher percentage of the organisms will be transported through the unit, because the biological active "living filter" has not been established. A rapid penetration can sometimes also be seen when a unit has been in use for a longer period, due to partially clogging and channeling. The importance of changes in ionic composition has been stressed. For a good reduction it is essential that sudden dosing of large volumes of water with a low ionic strength is avoided.

REFERENCES

1. The National Swedich Environmental Protection Board: 1971, "Skydd av vattentäkter." (Protection of drinking water sources.) 4 pp. 1-30.

2. Hutchinson, M.: 1974, "Microbial aspects of groundwater pollution." In: Groundwater pollution in Europe (ed. J.A. Cole) Port Washington, N.Y.

3. Ballangrud, P., O. Helle, A. Stavn & R. Kristiansen: 1981, "Fjerning av ulike organismetyper i jordrenseanlegg for avløpsvann. I. Effekter av ulike temperaturer og belastningsfrekvenser." (Reduction of different types of organisms in soil infiltration units for wastewater treatment. I. Effects of different temperatures and load intervals.) NLVF-report 2, pp. 39-59.

4. Kjellberg, S., M. Hermansson, T.A. Stenström & T. Korhonen: "Hydrophobic and electrostatic characterization of surface structures of bacteria and its relationship to adhesion to an air/water interface." Submitted for publication 1981.

5. Stenström, T.A. & S. Hoffner: 1979, "Patogeners och indikatororganismers överlevnad och transport i mark och grundvatten - Litteraturstudie." (Survival and transportation of microbial pathogens and indicators in soil and groundwater - A literature survey.) Project Report 1, 127 pp.

6. Stenström, T.A. & I. Blomén: 1981, "Tarmbakteriers överlevnad i mark - Laboratoriestudier. Bakteriestam, temperatur, jordtyp och inledande studier av markmikroflorans inverkan." (Survival of enteric bacteria in soil - Laboratory studies. Differences due to bacterial strain, temperature and type of soil.) Project report No. 4, 88 pp.

7. Andersson, M., F. Nyberg & T.A. Stenström: 1980, "Markbäddar och infiltrationsanläggningar. Rapport över verksamheten under 1979/1980." (Sandfilter trenches and infiltration units. Annual report 1979/1980.) TRITA-VAT 3810 88 pp.

REFERENCES

1. [illegible] ... [illegible] ... [illegible] Protection [illegible] ... [illegible] "[illegible] ... [illegible] for drinking water [illegible] ... [illegible]

2. [illegible] ... [illegible] "[illegible] aspects of groundwater pollution." In: Groundwater pollution in Europe (ed. [illegible]). [illegible], Washington, [illegible]

3. [illegible] ... [illegible] ... [illegible] ... [illegible] potential [illegible] ... [illegible] temperatures [illegible] ... [illegible] (reduction of different types of organisms in soil infiltration units for wastewater treatment. [illegible] ... [illegible] of different temperatures and load intervals. [illegible] report [illegible]

4. [illegible] ... [illegible] ... [illegible] characterization of [illegible] ... [illegible]

5. [illegible] ... [illegible] ... [illegible] transport [illegible] ... [illegible] in soil and groundwater [illegible] ... [illegible] Report [illegible] pp.

6. [illegible] ... [illegible]

SESSION V

SITE SELECTION CRITERIA FOR ON-SITE DISPOSAL SYSTEMS

Chairman: A.S. Eikum

EVALUATION OF SITE SUITABILITY FOR SUBSURFACE SOIL ABSORPTION OF WASTEWATER

Richard J. Otis, P.E.

Professor of Civil & Environmental Engineering
University of Wisconsin, Madison, Wisconsin 53706, USA

INTRODUCTION

The environment into which the wastewater is discharged, can be a valuable part of the onsite disposal system. If utilized properly, it can provide excellent treatment at little cost. However, if stressed beyond its assimilative capacity, the system will fail, resulting in a public health hazard or environmental degradation. Therefore, careful site evaluation is a crucial first step of onsite system design.

The primary criterion for selection of one design over another is preservation of public health while preventing environmental degradation. Secondary criteria are ease of operating and maintaining the system and cost. In general, systems designed to discharge partially treated wastewater to the soil for ultimate disposal are the most reliable and least costly. This is due to the soil's very large capacity to transform and recycle most pollutants found in domestic wastewaters, eliminating the need for extensive pretreatment. The assimilative capacity of some surface waters also may be great, but the quality of the wastewater to be discharged to them is usually specified by local water quality agencies. To achieve the specified quality may

A. S. Eikum and R. W. Seabloom (eds.), Alternative Wastewater Treatment, 185–198.

require a more complex and costly system. Evaporation requires little pretreatment, but this method of disposal is severely limited by local climatic conditions. Therefore, the soil should be carefully evaluated prior to the investigation of other receiving environments.

Studies have shown that 60 to 120 cm (2 to 4 ft) of unsaturated soil below the liquid/soil interface is sufficient to remove bacteria and viruses to acceptable levels, and nearly all the organics and phosphorus, while most of the nitrogen in the waste is quickly nitrified and is leached through the soil (1). Soils with rapid permeabilities may require greater unsaturated depths than soils with slow permeabilities.

Where soil and other site characteristics are unsuitable for conventional subsurface soil absorption systems, other subsurface soil absorption systems may be possible. Though these other systems may be more costly to construct than systems employing surface water discharge or evaporation, their reliable performance with a minimum of supervision may make them the preferred alternatives.

SITE EVALUATION STRATEGY

The objective of the site evaluation is to investigate the characteristics of the area for their potential to treat and dispose of wastewater. A good site evaluation is one that provides sufficient information to select the most appropriate treatment and disposal system from the range of feasible options. This requires that the site evaluation begin with all options in mind, eliminating infeasible options only as the collected site data indicate.

The site evaluation should be done in a logical and systematic manner to ensure the information collected is useful and in sufficient detail. A suggested strategy is outlined in

Table 1. It is based on the assumption that subsurface soil absorption is the most appropriate method of disposal. If the soils are found to be unsuitable, then the site's suitability for other disposal methods is evaluated.

Table 1. Suggested Site Evaluation Procedure.

Step	Data Collected
Client Contact	Location and description of lot Type of use Volume and characteristics of wastewater
Preliminary Evaluation	Available resource information (soil maps, geology, etc.) Records of onsite systems in surrounding area
Field Testing	Topography and landscape features Soil profile characteristics Hydraulic conductivity
Other Site Characteristics	If needed, site suitability for evaporation or discharge to surface waters should be evaluated
Organization of Field Information	Compilation of all data into usable form.

CLIENT CONTACT

As much information as possible concerning the lot and its intended use should be obtained from the client to help focus the site evaluation on the more important characteristics. This information would include the following:

Lot Description

- Location
- Size and shape
- General features

Wastewater Characterization

- Type

  domestic
  commercial
  industrial

- Volume

  daily average and peak flows
  seasonal variations
  future increases

- Important characteristics

  grease
  solids (settleable/nonsettleable)
  BOD (settleable/dissolved)
  hazardous and toxic substances
  other.

PRELIMINARY EVALUATION

Since the site characteristics constrain the method of disposal more than the other system components, the disposal method must be selected first. To do this properly, a detailed site evaluation is required. However, the site characteristics which must be evaluated in detail can vary with the method of disposal chosen. The purpose of this first step is to eliminate disposal options with the least potential so that the detailed site evaluation can concentrate on the more promising options.

Any available resource information concerning the site and the surrounding area should be gathered. This information may lack accuracy, but it can be useful in identifying potential problems or particular features to investigate. Sources of information include soil surveys, topographic and geologic maps and local records of the performance of existing systems in the area. A visual site survey with a hand auger is also valuable. From this information, the potentially feasible disposal options should be identified to help focus subsequent field testing. Table 2 can be used as a guide.

Table 2. Feasibility of Different Disposal Methods under Various Site Conditions (2).

Method	Site Constraints											
	Soil Permeability			Depth to Bedrock			Depth to Water table		Slope			Small Lot Size
	Very Rapid	Rapid-Moderate	Slow-Very Slow	Shallow and Porous	Shallow and Nonporous	Deep	Shallow	Deep	0-5%	5-15%	15%	
Trenches		X	X^2			X		X	X	X	X	X^4
Beds		X				X		X	X			X
Pits		X				X		X	X	X	X	X
Mounds	X	X	X	X	X	X	X	X	X	X		
Fill Systems	X	X^1	X^1	X	X	X	X	X	X	X	X	X^4
Sand-Lined Trenches or Beds	X	X	X^2			X		X	X	X^3	X^3	X^4
Artificially Drained Systems		X				X	X		X	X	X^3	
Evaporation Infiltration Lagoons		X	X^5			X		X	X			
Evaporation Lagoons (lined)[4,5]	X	X	X	X	X	X	X	X	X			
ET Beds or Trenches (lined)[4,5]	X	X	X	X	X	X	X	X	X	X^6		
ETA Beds or Trenches[4]		X	X			X		X	X	X	X	X

[1] Only where surface soil can be stripped to expose sand or sandy loam material.

[2] Construct only during dry soil conditions. Use trench configuration only.

[3] Trenches only.

[4] Flow reduction suggested.

[5] High Evaporation potential required.

[6] Recommended for south-facing slopes only.

X means system can function effectively with that constraint.

FIELD TESTING

Field testing begins with a visual survey of the parcel to locate potential sites for subsurface soil absorption. Detailed soils investigations will then be made at these sites. If no sites can be found from either the visual survey or detailed investigation, site suitability for evaporation or surface water discharge should be evaluated.

Visual Survey

A visual survey (preferably with a hand auger or soil probe) is made to locate the areas on the lot with the greatest potential for subsurface soil absorption. The following should be noted and marked on the plot plan:

General site features. The location of any depressions, gullies, steep slopes, rocks or rock outcrops, surface waters, roads, buildings and other obvious land and surface features should be noted and marked on the plot plan. Well travelled or compacted areas should be avoided.

Landscape position. Noting the landscape position and land form at the site is useful in estimating surface and subsurface drainage patterns. For example, ridge lines, hill tops and side slopes can be expected to have good surface and subsurface drainage, while depressions and foot slopes are more likely to be poorly drained. Figure 1 can be used as a guide for identifying landscape positions.

Flooding hazards. Areas of obvious flood hazard should be avoided. (If necessary, soil absorption systems may be installed in flood fringes out of the flood way.)

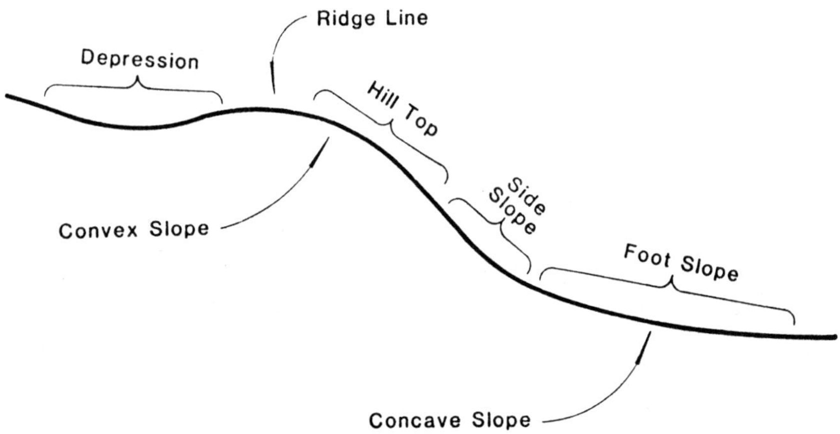

Figure 1. Landscape Positions.

Vegetation. The type and size of the existing vegetation should be noted. The depth and drainage characteristics of the soil often can be indicated from the type of vegetation. Large trees that must be removed or are to be saved may alter the layout of the system.

Slope. The type and degree of slope indicates surface drainage problems and areas to avoid because of construction problems. Concave slopes cause surface runoff to converge while convex slopes disperse the runoff. Slopes greater than 20 to 25 percent present difficulties to excavating equipment and some system designs. Abney or hand levels may be sufficient for simple slopes or small systems, but a topographic survey is necessary for all larger systems.

Horizontal setbacks. Setbacks from wells, surface waters, buildings, property lines, etc., should be maintained on the parcel and between neighboring parcels to minimize the threat to public health if a failure should occur. The setbacks required are usually specified by local codes.

Soil Borings

Detailed evaluation of soil characteristics is done in the areas selected during the visual survey. This can be done best from a pit excavated large enough to enter. However, an experienced soil tester can do a satisfactory job by using a hand auger or probe. Both methods are suggested. Hand tools can be used to determine soil variability over the area and pits used to describe in detail the various soils found. Power augers should not be used because the soil characteristics can be altered markedly.

Location, depth and number. Pits should be dug around the perimeter of the area. Pits dug within the absorption area often settle after the system is installed, disrupting the system. Hand augers can be used within the area. Pits should be oriented such that the sun hits one face directly for good color observation. The borings should be deep enough to insure that a sufficient depth of unsaturated soil exists.

Sufficient borings should be made to describe adequately the soils and their variability. Each should be carefully located in relation to a permanent bench mark. The ground surface elevation at each pit relative to the bench mark is also desirable.

Soil horizons. Any obvious soil horizons are tentatively identified from differences in color, texture or structure.

Soil texture. Beginning at the top or bottom of the pit sidewall, the texture of each horizon is identified. Hand texturing can be done by moistening a sample and working it until it has the consistence of putty. Using Table 2, the texture can be described quickly. When the textures have been determined for each layer, the depth, tickness and texture of each layer is recorded.

Table 2. Textural Properties of Mineral Soils (2).

Soil Class	Feeling and Appearance: Dry Soil	Feeling and Appearance: Moist Soil
Sand	Loose, single grains which feel gritty. Squeezed in the hand, the soil mass falls apart when the pressure is released.	Squeezed in the hand, it forms a cast which crumbles when touched. Does not form a ribbon between thumb and forefinger.
Sandy Loam	Aggregates easily crushed; very faint velvety feeling initially but with continued rubbing the gritty feeling of sand soon dominates.	Forms a cast which bears careful handling without breaking. Does not form a ribbon between thumb and forefinger.
Loam	Aggregates are crushed under moderate pressure; clods can be quite firm. When pulverized, loam has velvety feel that becomes gritty with continued rubbing. Casts bear careful handling.	Cast can be handled quite freely without breaking. Very slight tendency to ribbon between thumb and forefinger. Rubbed surface is rough.
Silt Loam	Aggregates are firm but may be crushed under moderate pressure. Clods are firm to hard. Smooth, flour-like feel dominates when soil is pulverized.	Cast can be freely handled without breaking. Slight tendency to ribbon between thumb and forefinger. Rubbed surface has a broken or rippled appearance.
Clay Loam	Very firm aggregates and hard clods that strongly resist crushing by hand. When pulverized, the soil takes on a somewhat gritty feeling due to the harshness of the very small aggregates which persist.	Cast can bear much handling without breaking. Pinched between the thumb and forefinger, it forms a ribbon whose surface tends to feel slightly gritty when dampened and rubbed. Soil is plastic, sticky and puddles easily.
Clay	Aggregates are hard; clods are extremely hard and strongly resist crushing by hand. When pulverized, it has a grit-like texture due to the harshness of numerous very small aggregates which persist.	Casts can bear considerable handling without breaking. Forms a flexible ribbon between thumb and forefinger and retains its plasticity when elongated. Rubbed surface has a very smooth, satin feeling. Sticky when wet and easily puddled.

Soil structure. The sidewall of the pit is carefully examined, using a pick, knife or similar device to expose the natural cleavages and planes of weakness. The durability of each structural unit is estimated by noting whether it withstands handling. If no cracks are visible, a sample of the soil is picked out and carefully separated into structural units by hand until any further breakdown can be achieved only by fracturing.

Soil color. Soil color and color patterns can be used to predict the natural soil drainage. Uniform red, yellow or brown colors indicate the soil is well-drained and seldom or never saturated with water, while gray or blue colors indicate that the soil is saturated for extended periods or continuously. Mottled soils are usually an indication that the soil is periodically saturated.

Bedrock. Bedrock may be in such a state of decay that it is difficult to determine where the true bedrock surface lies. It may be defined as that point where less than 50 percent of the excavated material is unconsolidated. The surface of sandstone bedrock can be defined as the point where resistance to penetration with a knife is encountered.

Bulk density. Relative bulk densities of each horizon can be detected by pushing a knife or other instrument into the soil. If one horizon offers considerably more resistance to penetration than others, its bulk density is probably higher. However, in some cases, cementing agents between soil grains or peds may be the cause of resistance.

Swelling clays. Swelling clays tend to be more sticky and plastic when wet.

Hydraulic Conductivity

In the areas where the soil borings indicate suitable soil for subsurface disposal, hydraulic conductivity testing follows.

Several methods of measuring the soil's ability to transmit water have been developed. The "percolation test" is the most commonly used. When run properly, it can give an approximate measure of the soil's saturated hydraulic conductivity. The most common test procedure is described in Table 3. Common errors made in running the test are poor hole preparation, inadequate soaking and inaccurate measurements.

Though highly variable and criticized for its inaccuracy, the test can be useful if used together with the soil boring data. If results from properly run percolation tests do not seem to agree with the texture of the soil, as shown in Table 4, then structure or mineralogy may be significant. Further investigations may be warranted.

Hydrogeologic Investigations

If the soil is to be used to dispose of large volumes of wastewater daily, then hydrogeologic investigations are necessary to determine if the soils have the capacity to conduct the liquid away from the infiltration area without becoming saturated to within 60 to 120 cm (2 to 4 ft) of the infiltration surface.

Groundwater elevation. The groundwater elevation and seasonal variations must be determined by monitoring wells and soil patterns in the soil profile. Soils with perched water table conditions should be avoided.

Groundwater gradient. Horisontal gradients are determined by measuring the water elevation in wells just penetrating the phreatic surface. Vertical gradients are determined from two or more wells at the same location, but at different depths within the water table.

Hydraulic conductivity. The hydraulic conductivity of the soil materials within the saturated zone is measured by pumping tests or permeameter tests in the laboratory. Both vertical and horizontal conductivities are desirable.

Table 3. Falling Head Percolation Test Procedure (2).

1. Number and Location of Tests

Commonly a minimum of three percolation tests are performed within the area proposed for an absorption system. They are spaced uniformly throughout the area. If soil conditions are highly variable, more tests may be required.

2. Preparation of Test Hole

The diameter of each test hole is 6 in., dug or bored to the proposed depths at the absorption systems or to the most limiting soil horizon. To expose a natural soil surface, the sides of the hole are scratched with a sharp pointed instrument and the loose material is removed from the bottom of the test hole. Two inches of 1/2 to 3/4 in. gravel are placed in the hole to protect the bottom from scouring action when the water is added.

3. Soaking Period

The hole is carefully filled with at least 12 in. of clear water. This depth of water should be maintained for at least 4 hr and preferably overnight if clay soils are present. A funnel with an attached hose or similar device may be used to prevent water from washing down the sides of the hole. Automatic siphons or float valves may be employed to automatically maintain the water level during the soaking period. It is extremely important that the soil be allowed to soak for a sufficiently long period of time to allow the soil to swell if accurate results are to be obtained.

In sandy soils with little or no clay, soaking is not necessary. If, after filling the hole twice with 12 in. of water, the water seeps completely away in less than ten minutes, the test can proceed immediately.

4. Measurement of the Percolation Rate

Except for sandy soils, percolation rate measurements are made 15 hr but no more than 30 hr after the soaking period began. Any soil that sloughed into the hole during the soaking period is removed and the water level is adjusted to 6 in. above the gravel (or 8 in. above the bottom of the hole). At no time during the test is the water level allowed to rise more than 6 in. above the gravel.

Immediately after adjustment, the water level is measured from a fixed reference point to the nearest 1/16 in. at 30 min intervals. The test is continued until two successive water level drops do not vary by more than 1/16 in. At least three measurements are made.

After each measurement, the water level is readjusted to the 6 in. level. The last water level drop is used to calculate the percolation rate.

In sandy soils or soils in which the first 6 in. of water added after the soaking period seeps away in less than 30 min, water level measurements are made at 10 min intervals for a 1 hr period. The last water level drop is used to calculate the percolation rate.

5. Calculation of the Percolation Rate

The percolation rate is calculated for each test hole by dividing the time interval used between measurements by the magnitude of the last water level drop. This calculation results in a percolation rate in terms of min/in. To determine the percolation rate for the area, the rates obtained from each hole are averaged. (If tests in the area vary by more than 20 min/in., variations in soil type are indicated. Under these circumstances, percolation rates should not be averaged.)

Example: If the last measured drop in water level after 30 min is 5/8 in., the percolation rate = (30 min)/(5/8 in.) = 48 min/in.

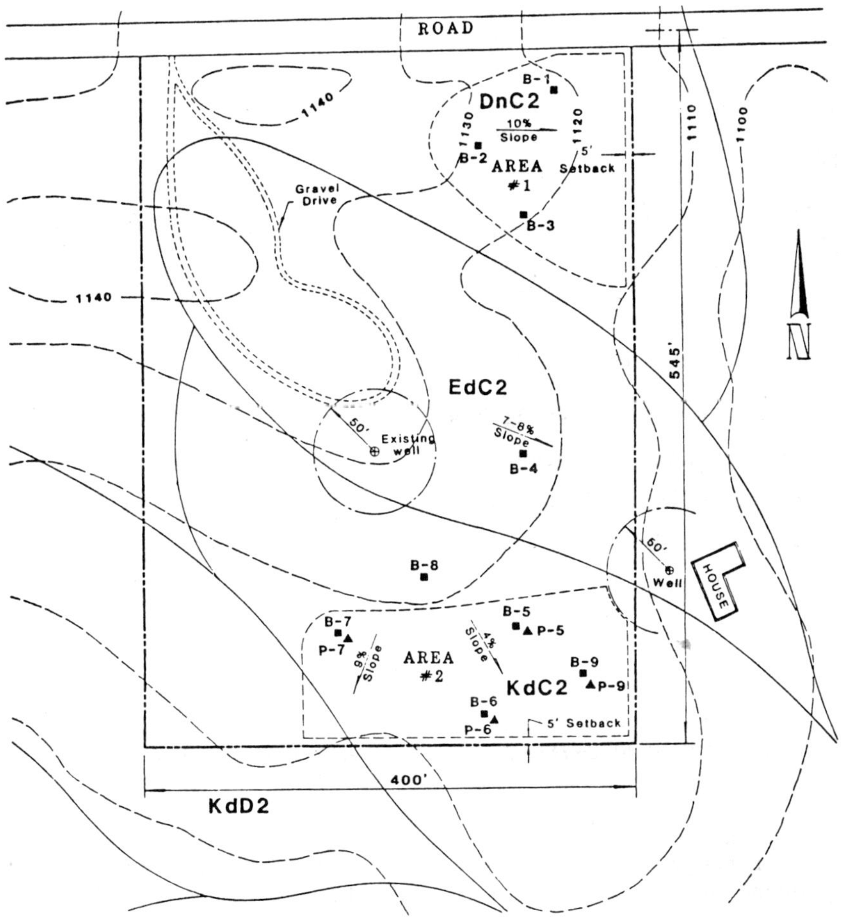

Figure 2. Completed Plot Plan of Lot, Showing Information Gathered from the Site Evaluation.

Table 4. Estimated Hydraulic Characteristics of Soil (3).

Soil Texture	Permeability		Percolation	
	cm/day	(in/hr)	min/cm	(min/in)
Sand	>360	(>6.0)	4	(<10)
Sandy loams) Porous silt loams) Silty clay loams)	12-360	(0.2-6.0)	4-18	(10-45)
Clays, compact) Silt loams) Silty clay loams)	<12	(<0.2)	>18	(>45)

ORGANIZING SITE INFORMATION

When the field testing has been completed, all relevant information should be organized and presented in a clear fashion. A good method is to draw a plot plan of the site, detailing all observations made. An example of a completed plot plan is illustrated in Figure 2.

REFERENCES

1. University of Wisconsin: 1978, Management of small waste flows. EPA-600/2-78-173. U.S. Environmental Protection Agency, Cincinnati, Ohio.

2. U.S. Environmental Protection Agency: 1980, Onsite wastewater treatment and disposal systems: design manual. Technology Transfer, Cincinnati, Ohio.

3. Bouma, J.: 1975, Unsaturated Flow During Soil Treatment of Septic Tank Effluent. J. Environ. Eng. Div., Am. Soc. Civ. Eng., 101; pp. 967-983.

SELECTION OF DISPOSAL SITES IN NORWAY

Petter D. Jenssen

Research Assistant, Agricultural Research Council
of Norway, The Agricultural University of Norway,
1432 Ås - NLH, Norway

INTRODUCTION

The guidelines for site selection are inadequate in the present Norwegian codes for onsite sewage disposal. Knowledge of the climate and sedimentologic and hydraulic properties of different soils are fundamental in developing site selection procedures.

The lack of skilled people to perform site investigations calls for simple criteria which nontheless have to be sufficient for predicting the soil absorption capabilities. Methods and criteria which are little influenced by the individual investigator are needed.

CLIMATE

The climate in Norway is mainly humid, but arctic in the northern part. The precipitation is highest in the west (Figure 1), where moist air from the North Sea is forced up to a higher altitude by the mountains and releases some of its moisture. The precipitation in some areas exceeds 2500 mm/year. In Norway, there are also some arid regions which have a mean annual precipitation

A. S. Eikum and R. W. Seabloom (eds.), Alternative Wastewater Treatment, 199–212.

less than 500 mm/year. In these regions there are locations where evaporation is higher than percolation and on silty soils there are problems with high soil salinity. Most of the country has an annual precipitation around 1000 mm/year. The monthly precipitation is highest in the fall in the western and northern parts. In the southeastern part the monthly precipitation is highest in the summer. The high precipitation in the fall together with melting of the snow in the spring creates two percolation peaks, one in the spring and one in the fall, where many soils are nearly saturated. Soil absorption systems have to be designed to function under such conditions. The humid climate of Norway also limits the use of systems, based solely on evapotranspiration.

SOILS

Most Norwegian sediments date back to the last glaciation, Weichselian, and the period shortly after this, the beginning of Holocene.

A simplified soil map of Norway is shown in Figure 2. About 90 percent of the land is covered by tills. Glaciofluvial and fluvial deposits are primarily located in the valleys.

Large areas of marine sediments, mainly silt and clays, are found in the southeastern part of the country (north and south of Oslo) and in mid Norway represented by the area around the Trondheim Fjord. Due to the high clay content marine sediments are generally considered unsuitable for conventional soil absorption systems.

There is a major difference between the soils of western Norway and the eastern part of southern Norway and Finmark. In the western part bedrock with a thinner and more dispersed cover of till dominates. In the eastern part along the Swedish border tills of greater thickness cover the land.

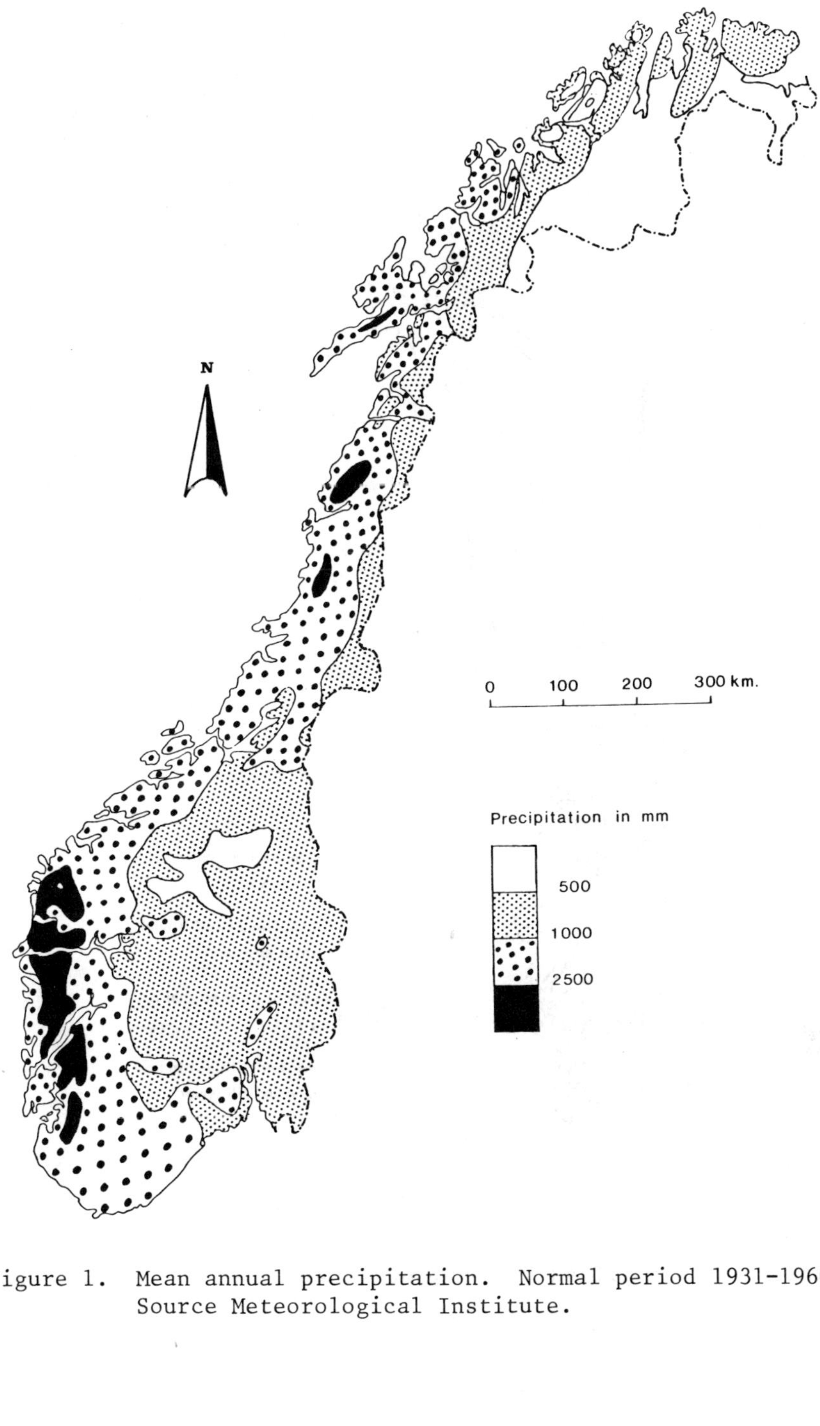

Figure 1. Mean annual precipitation. Normal period 1931-1960.
Source Meteorological Institute.

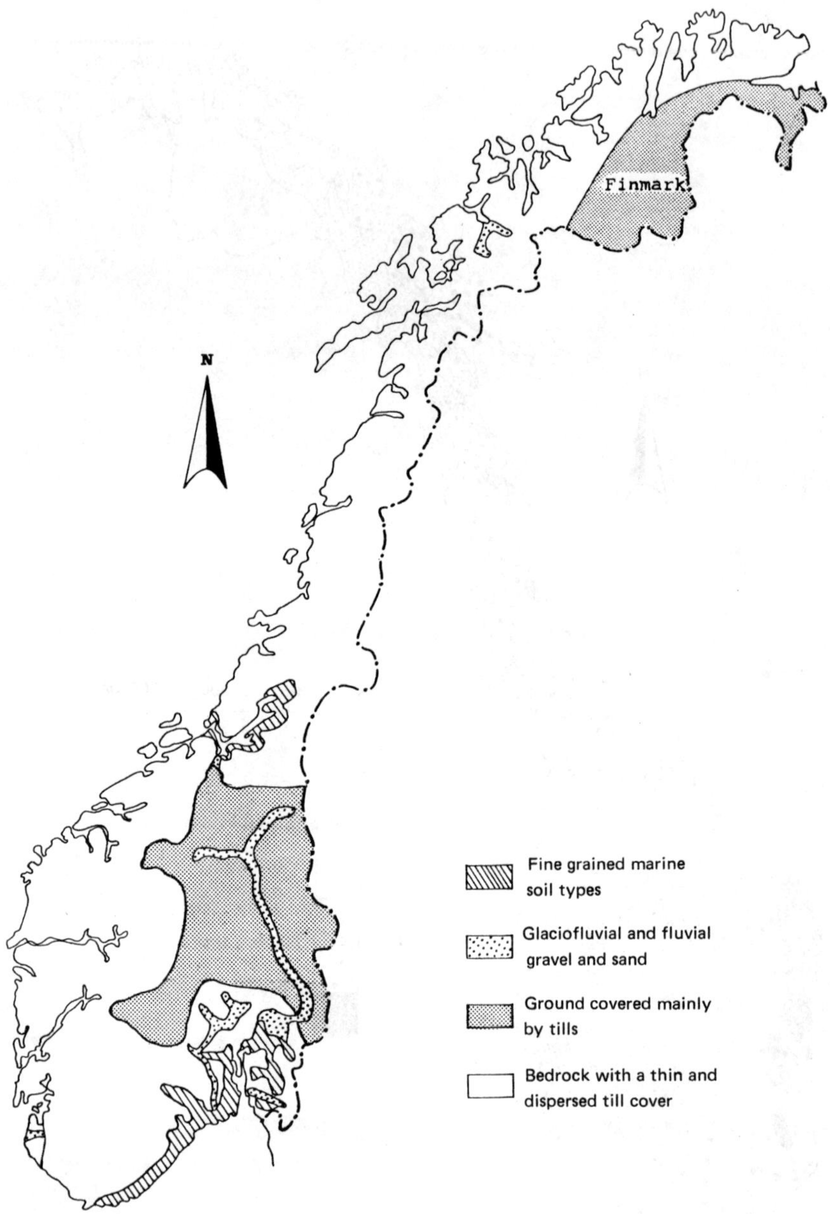

Figure 2. A simplified soil map of Norway.

The difference in soil depth between eastern and western Norway is mainly caused by the topography. Western Norway is characterized by high relief, narrow fjords and valleys surrounded by high mountains. The thin and dispersed till does not offer many good sites for absorption systems. The soils are thicker in depressions. In order to obtain sufficient soil depth for construction of conventional soil absorption systems, concave parts of the terrain have to be searched. Generally, this is not good practice, and problems with a high water table, converging flow and danger of polluted water reaching the surface downslope from the disposal site occur. In these areas there are few sites which have the capacity to absorb sewage effluent from more than one dwelling.

In the eastern part of south Norway and Finmark the topography is more level than in the western part of the country, hence the soil cover is thicker and more continuous. Till areas dominate. The average soil depth in the till areas is about 5 meters (2). Due to high compaction and poor sorting the hydraulic conductivity of the tills is generally low. Temporary high groundwater in the rainy seasons is also common. This is caused by the relatively permeable A- and B-horizons while the C-horizon, which is not to the same extent exposed to freezing and biological activity, often is nearly impermeable. In the till areas absorption of septic tank effluent is normally controlled by the hydraulic properties of the soil system surrounding the leaching field. In general till areas produce few good sites for absorption of sewage effluent, but like other Norwegian soil types, tills are heterogeneous. By thorough field investigations it is possible to find sites where larger volumes of effluent can be absorbed.

Figure 3 shows the soil distribution of a valley representative for the inland parts of Norway.

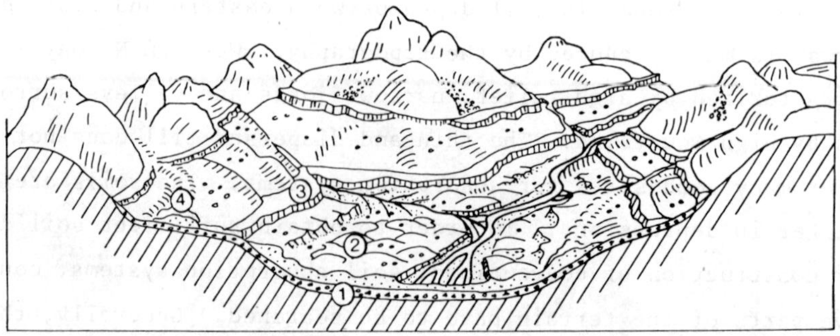

1. Basal till
2. Ablation till
3. Glaciofluvial terrace
4. Esker

Figure 3. Soil distribution of an inland valley in Norway.

Basal till covers the rock in the entire profile. This till has a high silt content and a high density. Hilly terrain in the valley bottom is often covered by ablation till. The ablation till generally has a low content of silt and finer material and it is relatively well sorted. These areas provide many potentially good sites on concave parts of the terrain. In fact, this topography and soil produce a kind of natural mounds. The hydraulic capacity for wastewater absorption in such areas is much dependent upon whether the sediment underneath the ablation till is basal till or glaciofluvial deposits.

Fluvial deposits in the valley bottoms are normally well graded and might provide good sites. But Norway is a mountainous country and many rivers are swift flowing. The fluvial sediments are therefore often very coarse. Other problems in such sediments are thin layers of fine material which can produce hydraulic barriers, and the risk of flooding or seasonally high groundwater.

The glaciofluvial deposits often form terraces in the valley sides, or eskers. These sediments are mainly composed of well sorted sand and gravel. They are well drained, and the groundwater level is normally low. The glaciofluvial deposits therefore provide some of the best sites for absorption systems. Although glaciofluvial sediments are fairly homogeneous and well sorted from a Norwegian point of view, great horizontal and vertical variation occurs.

In fact, heterogeneity is characteristic for all types of Norwegian sediments. Compared to the soils further south in Europe, Norwegian soils are coarse and poorly sorted, and the soil cover is often dispersed and extremely thin.

Knowledge of the sedimentological properties of soils of different genesis combined with quarternary maps can aid site selection. Quarternary maps are with few exceptions presented on the scale 1:50,000. Due to soil heterogeneity these maps are only suited for finding potential sites. Soil maps on the scale 1:5,000 which give information about texture, drainage class, layering and slope are still on the experimental level. Field investigations are therefore necessary to confirm interpretations of the maps.

PRESENT REQUIREMENTS

At present there are no manuals to guide field investigations. Manuals are under preparation and will be presented in 1982. In the present requirements a sieve analysis for determination of the grain size distribution is the main criterion for site selection and evaluation. The other criterion of importance with regard to evaluation of the soil systems, is the distance to the groundwater table which must be more than 0.8 m from the bottom of the absorption trench.

The grain size distribution curve is used to determine whether soil absorption is possible or not, and to size the absorption system. If the grain size distribution curve lies entirely within part A or part B in the diagram shown in Figure 4, absorption is possible. If any part of the curve lies outside part A or B, absorption systems are generally not permitted. The loading of the system is determined from the position of the curve, whether it is in part A or B or both. For single dwellings there are two standard lengths of the absorption trench (absorption beds are not mentioned), 20 or 40 m with a bottom width of 0.8 m.

The grain size distribution curve shown in Figure 4 is from a till in south-eastern Norway. From the position of the curve it can be determined that absorption is possible. Five conventional percolation tests were conducted in this area. The highest percolation rate was 24 min/cm (60 min/inch). The lowest showed no percolation at all, hence the soil was nearly impermeable.

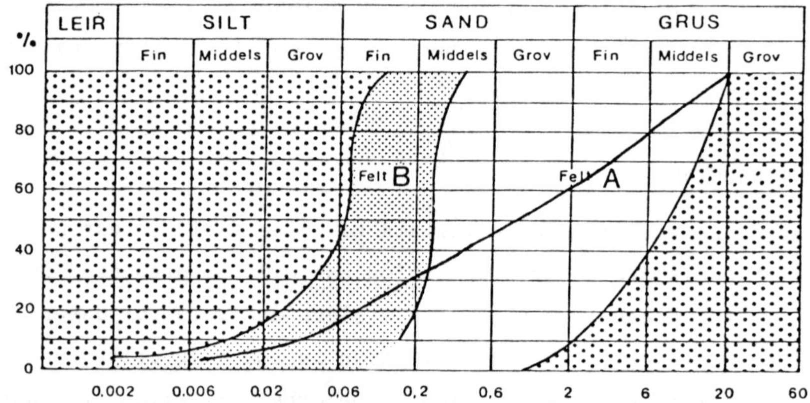

Figure 4. Grain size distribution diagram for determination of absorption possibilities and sizing of absorption systems.

This example shows that the texture alone does not give enough information about the hydraulic properties of a soil, especially when it is poorly sorted and highly compacted which is the case in most Norwegian tills.

Percolation tests are described in our codes, but a percolation test is not required and therefore seldom used.

A serious problem in Norway is the lack of skilled people to perform site investigations. The future criteria for site selection must therefore be simple. The present criteria are simple, but far from adequate.

IMPROVED METHODS FOR MEASUREMENT OF THE PERCOLATION RATE AND HYDRAULIC CONDUCTIVITY

In the new manuals more emphasis will probably be put on percolation tests and measurements of hydraulic conductivity. During the past year several methods for measuring the percolation rate and hydraulic conductivity have been evaluated. The only method which it was possible to use in the field on all types of Norwegian sediments, was the percolation test.

But the validity of the percolation test has been questioned. According to Machmeier (4) one of the problems is the many different techniques used by the individuals who perform the test. To eliminate this it is suggested by Peterson (5) to run the test with a constant head instead of a falling head. Bouma (1) also found that the constant head percolation test is to be preferred over the falling head procedure because of lower variability.

In our studies we constructed a Mariotte cylinder to keep a constant water level in the test pit (Figure 5). The cylinder is graded and the volume flow per unit time is measured directly on the cylinder It is a problem to keep the walls of the test pit undisturbed during the test. Peterson (5) suggested lining

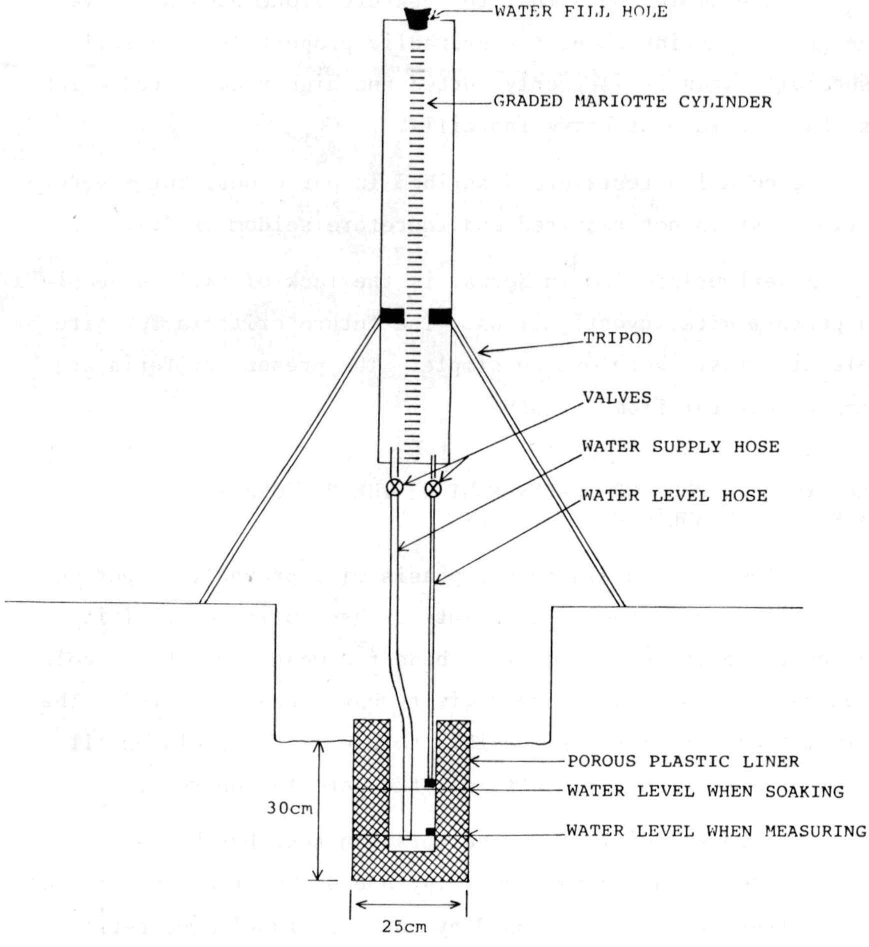

Figure 5. Mariotte cylinder and porous plastic pad used in percolation tests.

the pit with gravel. As gravel is not always easily accessable and heavy to transport, the pits were lined with porous plastic pads. The pads give good support to the walls and they reduce turbulence when filling the pit.

Different sizes of the test pit have been tried. The reasoning was that a larger area would give a more representative result. But the large pits often consumed very much water. For practical reasons the pit size therefore was reduced from 60x60 cm to 25x25 cm. A single piece of porous plastic is then used to line the pit.

Experience with this equipment showed that the percolation test was easier to run and the result depended more upon the soil conditions than upon the individual performing the test.

Layering is especially characteristic of Norwegian fluvial and glaciofluvial deposits. Such layers can restrict flow and thereby reduce the hydraulic capacity of the soil system. A simple tool for measuring the hydraulic conductivity in critical parts of the soil system is therefore needed. The use of small steel cylinders for measuring the hydraulic conductivity on undisturbed soil samples is proposed by Healy and Laak (3). Coring tools (Figure 6) and a measuring box (Figure 7) for such cylinders have been made.

To reduce possible flow along the soil/steel interface the cylinders are greased inside with vaseline before they are inserted into the soil. The measuring box is filled with water before the steel cylinders with the soil samples are fitted into it. The water is then forced up through the soil, thereby removing the air. When the sample is soaked, a small Mariotte cylinder is placed on top of the sample, the water level plug is removed and a head difference is created, i.e. the water level inside the cylinder is higher than the water level in the box. Measurements of the hydraulic conductivity is then obtained by the constant head method.

The author has compared this method with expensive lab equipment for measuring hydraulic conductivity on undisturbed

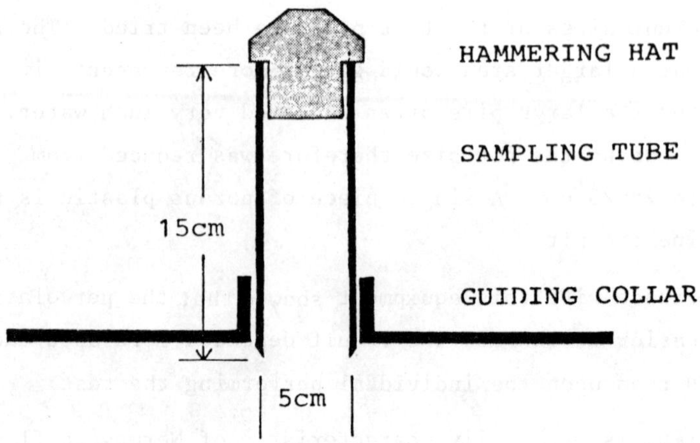

Figure 6. Coring tools and steel cylinder for soil sampling.

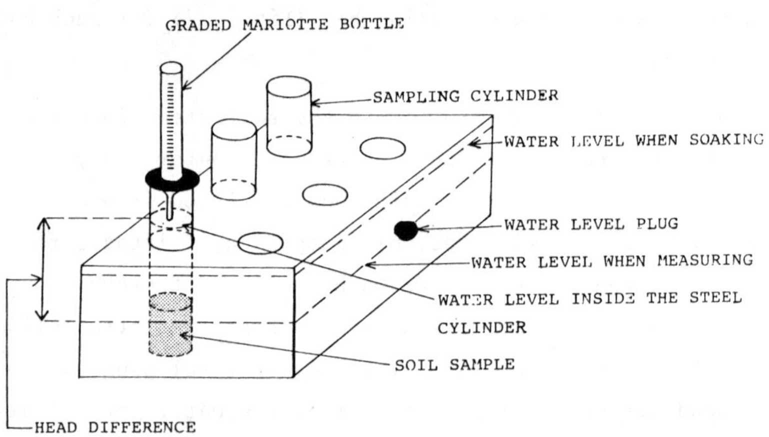

Figure 7. Field kit for measurement of hydraulic conductivity.

samples in small steel cylinders. The variabilities of the methods were much the same, but the lab equipment was more laborious.

SUMMARY

Norwegian climate is partly humid, partly arctic. This limits the use of systems based only on evapotranspiration. Many soils are nearly saturated in the spring and in the fall, and absorption systems have to be constructed to function under such conditions.

Norwegian soils are characterized by poor sorting and heterogeneity, and compared to the rest of Europe they are also extremely thin. About 90 percent of the country is covered by till. The till is mostly thin and dispersed and often unsuited for conventional absorption systems. The glaciofluvial deposits provide some of the best sites for absorption systems.

Due to the heterogeneity of the soils, field investigations have to be carried out. The lack of skilled people to perform the investigations calls for simple criteria for site selection and evaluation.

In the present code the main requirement is a sieve analysis. This is inadequate. In the new codes, which are under preparation, more emphasis will be put on direct measurements of the hydraulic properties. Equipment which makes the percolation test more accurate and less dependent on the individual investigator, is described. An improved method for measurement of the hydraulic conductivity in field is developed.

REFERENCES

1. Bouma, J.: 1971, "Evaluation of the field percolation test and an alternative procedure to test soil potential for disposal of septic tank effluent." Soil Sci. Soc. Amer. Proc. 35, pp. 871-875.

2. Haldorsen, S.: 1981, "Characteristics and genesis of Norwegian tills." In: Ehlers, J. (Ed.): Glacial deposits in northern Europe. (In print.)

3. Healy, K.A. & R. Laak: 1973, "Factors affecting the percolation test." J. Water Pollut. Control Fed. 45, pp.1508-1516.

4. Machmeier, R.E.: 1975, "Design criteria for soil treatment systems." Amer. Soc. Agr. Eng. Paper No. 75-2577, p.35.

5. Peterson, M.E.: 1980, "Soil percolation tests." J. Environ. Health 42, pp. 182-186.

SESSION VI

DESIGN, CONSTRUCTION, AND FUNCTION OF SEPTIC TANK SOIL ABSORPTION SYSTEMS

Chairman: A.S. Eikum

SUBSURFACE SOIL ABSORPTION SYSTEMS USED IN THE UNITED STATES

Richard J. Otis, P.E.

Professor of Civil & Environmental Engineering
University of Wisconsin, Madison, Wisconsin 53706, USA

INTRODUCTION

Methods of onsite wastewater treatment and disposal which utilize the soil for ultimate disposal of the liquid phase, must rely on the soil matrix to absorb and purify the wastewater. If the soil is not sufficiently permeable to absorb all the wastewater, surface seepage will occur, creating nuisances, public health hazards and surface water contamination. If the soil is too rapidly permeable or the unsaturated thickness of the soil is too thin, groundwater contamination will occur, leading to unsafe water supplies. Therefore, the soils must meet strict criteria if both absorption and purification are to be achieved.

Conventional trench and bed systems used in the United States are constructed in the natural soil. To be acceptable, the soil must have a percolation rate between 0.4 and 24 min/cm (1 and 60 min/in), and an unsaturated thickness below the infiltrative surface of 90 to 120 cm (3 to 4 ft). Less than 35 percent of all soils in the United States meet these criteria. Because there is public pressure to develop areas beyond the reach of municipal sewers where soils are unsuitable for conventional onsite systems,

A. S. Eikum and R. W. Seabloom (eds.), Alternative Wastewater Treatment, 215–234.

alternative designs have been developed. These are designed to provide the necessary soil conditions. This paper will discuss conventional trenches and beds, and mounds only.

CONVENTIONAL TRENCHES AND BEDS

Description

Trenches and beds are shallow, level excavations. The bottom of the excavation is filled with 15 cm (6 in) or more of clean rock or gravel over which the wastewater distribution piping is laid. Additional rock or gravel is placed over the pipe and covered with a semi-permeable barrier to prevent the backfill from penetrating the rock.

Trenches are narrow excavations usually 30 to 90 cm (1 to 3 ft) in width and up to 30.8 m (100 ft) or more in length. Only one distribution pipe per trench is used. The primary infiltration surfaces are the bottom and sidewalls of the excavation (see Figure 1). Beds are usually greater than 90 cm (3 ft) in width. More than one distribution pipe is used per bed. The primary infiltrative surface is the bottom of the excavation. Single beds are usually used, although dual beds loaded alternately are frequently constructed.

Site Criteria

Site criteria for trench and bed systems are summarized in Table 1. They are based on factors necessary to maintain reasonable infiltration rates and adequate treatment over a reasonable lifetime.

Design

The objectives in the design of subsurface soil absorption systems are to (1) ensure soil infiltration rates equal to or greater than the wastewater application rates, (2) ensure adequate

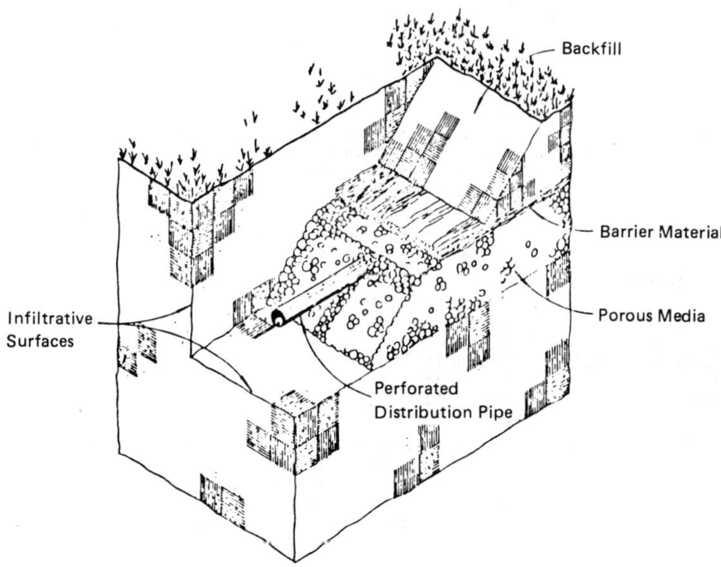

Figure 1. Schematic of a Typical Trench Design.

treatment of the wastewater is maintained within the soil, and (3) maintain these first two objectives over a long useful life.

Sizing the Infiltrative Surface

The design of a soil absorption system begins at the infiltrative surface where the wastewater enters the soil. With continued application of wastewater, a clogging mat forms on this surface, reducing the rate of wastewater infiltration below the percolative capacity of the surrounding soil. Therefore, the infiltrative surface must be sized on the basis of the expected infiltrative capacity of the clogging mat and the estimated daily wastewater flow.

The rate of infiltration through the clogging mat is dependent upon (1) the resistance of the clogging mat, (2) the hydrostatic pressure of the water ponded above, and (3) the matric potential of the underlying soil. Equilibrium infiltration rates

Table 1. Site Criteria for Trench and Bed Systems (U.S. EPA, 1980).

Landscape Position[a]	Level, well drained areas, crests of slopes, convex slopes most desirable. Avoid depressions, bases of slopes and concave slopes unless suitable surface drainage is provided.
Slope[a]	0 to 25%. Slopes in excess of 25% can be utilized but the use of construction machinery may be limited. Bed systems are limited to 0 to 5%.
Typical Horizontal Separation Distances[b]	
Water Supply Wells	50 - 100 ft
Surface Waters, Springs	50 - 100 ft
Escarpments, Manmade Cuts	10 - 20 ft
Boundary of Property	5 - 10 ft
Building Foundations	10 - 20 ft
Soil	
Texture	Soils with sandy or loamy textures are best suited. Gravelly and cobbley soils with open pores and slowly permeable clay soils are less desirable.
Structure	Strong granular, blocky or prismatic structures are desirable. Platy or unstructured massive soils should be avoided.
Color	Bright uniform colors indicate well-drained, well-aerated soils. Dull, gray or mottled soils indicate continuous or seasonal saturation and are unsuitable.
Layering	Soils exhibiting layers with distinct textural or structural changes should be carefully evaluated to insure water movement will not be severely restricted.
Unsaturated Depth	2 to 4 ft of unsaturated soil should exist between the bottom of the system and the seasonally high water table or bedrock.
Percolation Rate	1-60 min/in. (average of at least 3 percolation tests).[c] Systems can be constructed in soils with slower percolation rates, but soil damage during construction must be avoided.

[a] Landscape position and slope are more restrictive for beds because of the depths of cut on the upslope side.

[b] Intended only as a guide. Safe distance varies from site to site, based upon topography, soil permeability, ground water gradients, geology, etc.

[c] Soils with percolation rates <1 min/in. can be used for trenches and beds if the soil is replaced with a suitably thick (>2 ft) layer of loamy sand or sand.

through clogged soil surfaces have been measured (USPHS, 1967; Healy and Laak, 1973; Bouma, 1975). They vary from soil to soil, but often can be correlated with texture as shown in Table 2. However, soil structure, mineralogy, bulk density and other factors may affect the rates shown significantly.

To determine the infiltrative surface area, the estimated peak daily wastewater volume is divided by the loading rate selected. For single family home systems, the peak volume is

Table 2. Recommended Septic Tank Effluent Loading Rates for Trenches and Beds (U.S. EPA, 1980).

Soil Texture	Percolation Rate		Bottom Area Application Rate[1]	
	min/cm	(min/in)	cm/day	(gpd/ft^2)
Gravel, coarse sand	<0.4	(<1)	Not suitable	
Coarse to medium sand	0.4-2	(1-5)	5	(1.2)
Fine sand, loamy sand	2-6	(5-15)	3	(0.8)
Sandy loam, loam	6-12	(15-30)	2.5	(0.6)
Loam, porous silt loam	12-24	(30-60)	2	(0.45)
Silty clay loam, clay loam[2]	24-48	(60-120)	1	(0.2)[3]

1) May be suitable estimates for sidewall infiltration rates.

2) Soils without significant amounts of expandable clays.

3) Soils easily damaged during construction.

---

estimated from the size of the home. In the U.S., a flow of 570 l/d (150 gpd) per bedroom is used. This is based on the assumption that there are two people per bedroom, each producing 285 l/d (75 gpd).

Geometry of the Infiltrative Surface

The geometry of the infiltrative surface can affect the rate and degree of soil clogging, the extent of soil damage during construction, the soil horizons used for absorption and other factors.

Bottom vs. sidewall area. Both the horizontal bottom area and the vertical sidewalls of the excavation can act as infiltrative surfaces; however, the bottom must be ponded before the sidewall becomes an infiltrative surface. The resistance of the clogging mats and the hydraulic gradients at the bottom and sidewalls are not likely to be the same and, therefore, infiltration rates may be different. Clogging at the sidewall is likely to be less severe than at the bottom because suspended solids settle to the bottom and the rising and falling liquid levels cause dosing and resting cycles. The hydraulic gradient is less across the sidewall because the gravity potential is zero and the hydrostatic pressure is less significant than at the bottom. The objective in design is to maximize the area of the surface expected to have the highest infiltration rate. This is difficult to predict. Because of this difficulty, it is recommended that the loading rate used for the bottom area be the maximum rate applied to the sidewalls. In humid regions where percolating rain water reduces the matric potential along the sidewall, the bottom area should be the principal infiltrative surface. In dry regions, the sidewall can be used to a greater extent because the matric potential remains high at the sidewall.

Trench vs. bed design. Trenches are usually more desirable than beds because: (1) trenches can provide up to 5 times more sidewall area than beds for identical bottom areas, (2) less soil damage occurs during construction because the excavation equipment can straddle the trenches, eliminating the need to drive on the infiltrative surface, and (3) trenches can follow the contours on sloping sites to maintain the infiltrative surface in the same soil horizon and to keep excavation at a minimum. Beds may be acceptable where the soils are sands and loamy sands and the site is relatively level. These soils are not damaged as easily during construction (Univ. of Wis., 1978).

Shallow vs. deep designs. Shallow soil absorption systems have the following advantages over deep systems: (1) the upper soil horizons are usually more permeable than the deeper subsoil because of greater plant and soil fauna activity and less clay due to eluviation, (2) evapotranspiration is greater, (3) the upper horizons dry quicker than the subsoil, reducing construction delays, and (4) less excavation is necessary, reducing the cost. If kept in continuous operation, freezing is not a problem if there is 15 to 30 cm (6 to 12 in) of soil cover (Machmeier, 1981; Univ. of Wis., 1978).

Deep soil absorption systems have the following advantages over shallow systems: (1) increased depths permit increased sidewall area for the same bottom area, (2) a greater depth of liquid ponding increases the hydraulic gradient across the clogging mat, and (3) more permeable soil may exist at greater depths. However, in most instances, shallow systems are preferred.

Alternating fields. Dividing the absorption system into two or more fields to permit alternate use of each field may extend the life of the system. Common practice is to switch between fields on an annual basis. The field taken out of service is allowed to "rest" so that the infiltrative surface can be rejuvenated naturally through biodegradation of the clogging mat. The "resting" field also acts as a standby unit that can be put into immediate service if a failure occurs in another part of the system. Small alternating systems serving one or two homes are commonly divided into two equally sized fields, each containing 75 to 100 percent of the required infiltration surface area.

Wastewater Distribution

The method used to distribute the wastewater over the infiltrative surface can affect the degree of clogging and treatment performance of the system. Selection of the most appropriate method is based on the site characteristics.

Methods of distribution. There are three basic methods of wastewater distribution. Each has its own advantages and disadvantages (Otis et al., 1978a):

- Gravity flow:

Wastewater is allowed to flow by gravity through large diameter pipe directly from the treatment unit as wastewater is generated. It is characterized as "trickle flow". It is the simplest method and, therefore, the most commonly used. Distribution is very uneven, resulting in localized overloading and clogging which eventually progresses throughout the system (Converse, 1974; Univ. of Wis., 1978). Ultimately, continuous ponding of the absorption system occurs. This submerges the sidewalls, increasing the infiltration area and hydraulic gradient across the clogging mat. However, these benefits may be offset by more severe clogging (Bendixen et al., 1950; Winneberger et al., 1960; Thomas et al., 1966; Univ. of Wis., 1978). In very rapidly permeable soils, this is a poorly suited method because a clogging mat sufficient to provide treatment at the sites of localized loading may not develop (Univ. of Wis., 1978).

- Dosing:

Wastewater is stored after treatment for periodic discharge into the soil absorption field through large diameter perforated pipe by a pump, siphon or other device. The wastewater is distributed over a larger portion of the absorption system during each dose and the period between doses allows the infiltrative surface to drain, exposing the clogging mat to air. The cycles of alternate dosing and resting seem to promote degradation of the clogging mat. This appears to maintain higher infiltration rates and to extend the life of the system, but research evidence is conflicting (Otis et al., 1978b). In sand or coarser textured materials, rapid infiltration rates can lead to bacterial and viral contamination of shallow groundwater. Therefore, systems in coarse, granular soils should be dosed more frequently with small volumes of wastewater. In finer textured soils where clogging is more of a concern than treatment, large less-frequent doses are more suitable. See Table 3 for suggested dosing frequencies.

Table 3. Dosing Frequencies for Different Soil Textures (U.S. EPA, 1980).

Soil Texture	Dosing Frequency doses/day
Sand	4
Sandy loam	1
Loam	Not critical[1)]
Silt loam; silty clay loam	1
Clay	Not critical[1)]

1) Long-term resting provided by alternating fields may be desirable.

- Uniform application:

Wastewater is stored after treatment and periodically dosed uniformly over the entire bottom area of the system through a pressurized distribution network. In this manner, the wastewater can be applied uniformly throughout the entire system. The rate of application is below the saturated hydraulic conductivity of the soil so that unsaturated soil conditions are maintained, providing adequate treatment at all times. Localized overloading problems are avoided. Clogging also seems to be less severe (Univ. of Wis., 1978). See Table 3 for dosing frequencies.

Selection of distribution method. The selection of an appropriate method of distribution depends on whether improved absorption or improved treatment is the objective. This is determined by the permeability of the soil and the geometry of the infiltrative surface. Under some conditions, the method of application is not critical, so selection is based on simplicity and cost. Methods of distribution for various soils and geometries are listed in order of preference in Table 4. The design of different distribution networks is described by Otis et al., 1978a; Otis, 1981; and U.S. EPA, 1980.

Table 4. Recommended Methods of Effluent Distribution for Various System Geometries and Soil Permeabilities[1] (Otis, 1981).

Soil Permeability (Percolation Rate)	Trenches or Beds on Level Site	Multiple Trenches on Sloping Sites (>5%)
Very Rapid[2]	Uniform application[3]	Uniform application
<1 min/in (<0.04 cm/sec)	Dosing	Gravity
	Gravity	Dosing
Rapid	Uniform application	Gravity
1-10 min/in (4-0.4 cm/sec x 10^{-2})	Dosing	Uniform application
	Gravity	Dosing
Moderate	Dosing	Gravity
11-60 min/in (4-0.7 cm/sec x 10^{-3})	Gravity	Uniform application
	Uniform application[4]	Dosing
Slow	Not Critical	Not Critical
60 min/in (>0.7 x 10^{-3} cm/sec)	Uniform application[4]	

[1] Methods of application are listed in order of preference.

[2] Conventional soil absorption systems not recommended for these soils.

[3] Should be used exclusively in alternating field systems to ensure adequate treatment

[4] Preferred method for large flows.

MOUNDS

Description

A mound system is a subsurface soil absorption system that is elevated above the natural soil surface in a suitable fill material. Trenches or beds are constructed in the fill, maintaining one to two feet of fill material between the bottom of the seepage area and the natural soil. Septic tank effluent is pumped or siphoned into the gravel area through a pressure distribution network. The system is covered with a finer textured soil material. See Figure 2.

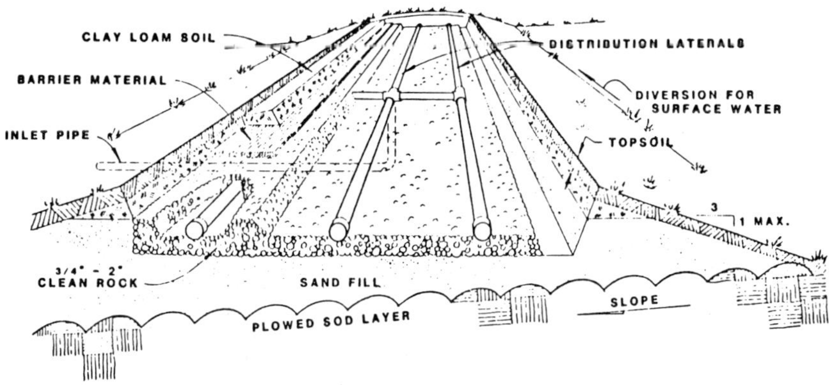

Figure 2. Mound System Schematic.

In slowly permeable soils, the primary function of the mound is absorption of the wastewater into the natural soil. By elevating the seepage area above the natural soil surface, several advantages result:

Absorption into the topsoil. The topsoil is usually more permeable than the subsoil because of greater soil flora and fauna activity. Conventional systems are installed below the topsoil, losing the benefit of the greater permeability. Once into the topsoil, wastewater can move laterally until absorbed by the subsoil. See Figure 3. (Transpiration by plants may also

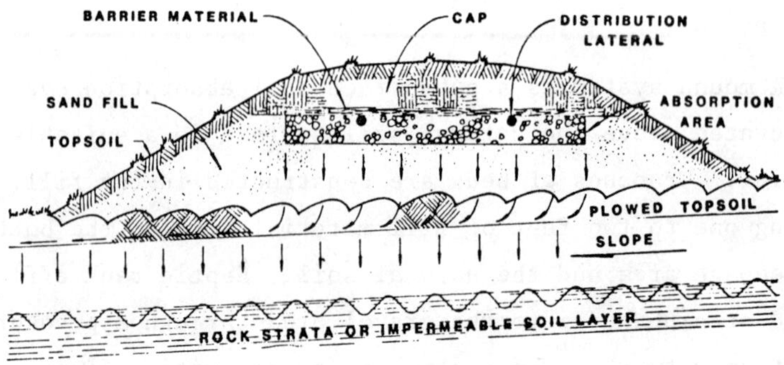

(a)

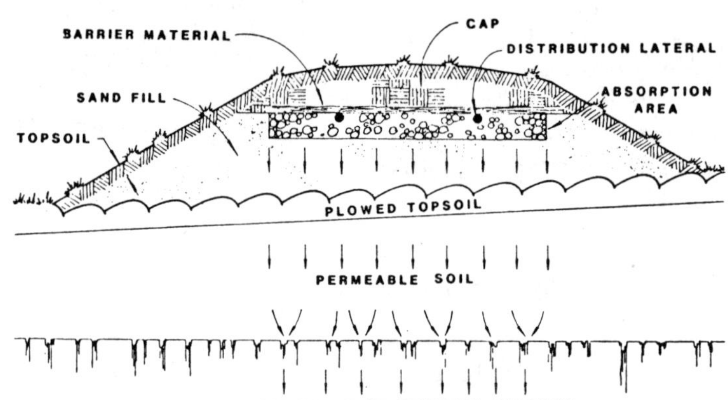

(b)

Figure 3. Schematic of Water Movement from a Mound System: (a) Slowly Permeable Soils; (b) Permeable Soils over High Water Tables or Shallow Porous Bedrock.

remove significant amounts of water during the growing season, but the mound is designed to function solely by absorption.)

Less restrictive clogging mat. Pressure distribution appears to reduce the severity of the clogging mat in coarse, textured soils such as the sandy fill material.

Construction damage minimized. Smearing and compaction of the wetter subsoil during construction is avoided since excavation in the natural soil is not necessary.

In shallow permeable soils, the primary function of a mound is to provide additional unsaturated soil material to adequately treat the wastewater before it reaches the groundwater. See Figure 3.

Several years of monitoring of laboratory models and full-scale field systems have demonstrated that mound systems consistently remove all waste contaminates of concern except for nitrogen (Univ. of Wis., 1978). There is some evidence that some nitrogen removal does occur in mound systems (Harkin et al., 1979). To maintain this treatment level, approximately 2 ft of natural unsaturated soil is required below the fill material.

Site Criteria

Mounds are used in slowly permeable soils and permeable soils with shallow water tables or shallow creviced or porous bedrock where conventional trench or bed systems are unsuitable. Site criteria for mounds are summarized in Table 5. These represent current practice for small systems and can be expected to become broader as experience is gained. Larger systems require more detailed hydrogeologic site investigations.

Design

Mound design is based on: (1) the estimated peak daily wastewater volume, (2) the fill characteristics, and (3) the

Table 5. Site Criteria for Mound Systems (U.S. EPA, 1980).

Item	Criteria
Landscape Position	Well-drained areas, level or sloping. Crests of slopes or convex slopes most desirable. Avoid depressions, bases of slopes and concave slopes unless suitable drainage is provided.
Slope	0 to 6% for soils with percolation rates slower than 60 min/in.[a]
	0 to 12% for soils with percolation rates faster than 60 min/in.[a]
Typical Horizontal Separation Distances from Edge of Basal Area	
Water Supply Wells	50 to 100 ft
Surface Waters, Springs	50 to 100 ft
Escarpments	10 to 20 ft
Boundary of Property	5 to 10 ft
Building Foundations	10 to 20 ft (30 ft when located upslope from a building in slowly permeable soils).
Soil	
Profile Description	Soils with a well-developed and relatively undisturbed A horizon (topsoil) are preferable. Old filled areas should be carefully investigated for abrupt textural changes that would affect water movement. Newly filled areas should be avoided until proper settlement occurs.
Unsaturated Depth	20 to 24 in. of unsaturated soil should exist between the original soil surface and seasonally saturated horizons or pervious or creviced bedrock.
Depth to Impermeable Barrier	3 to 5 ft[b]
Percolation Rate	0 to 120 min/in. measured at 12 to 20 in.[c]

[a] These are present limits used in Wisconsin, established to coincide with slope classes used by the Soil Conservation Service in soil mapping. Mounds have been sited on slopes greater than these, but experience is limited.

[b] Acceptable depth is site dependent.

[c] Tests are run at 20 in. unless water table is at 20 in., in which case test is run at 16 in. In shallow soils over pervious or creviced bedrock, tests are run at 12 in.

natural soil characteristics. Mounds must be designed to accept the peak daily wastewater flow without surface seepage or encroachment of the zone of saturation into the fill material.

For systems to serve single family residences, the peak volume is estimated from the size of the home, typically by the number of bedrooms. Common practice is to use 570 l/d (150 gpd) per bedroom.

The fill material must be selected before sizing of the mound can be done because the material's infiltrative capacity determines the required absorption bed area. Medium textured sands, sandy loams, soil mixtures, bottom ash, strip mine spoil and slags are being used or are being tested (Converse et al., 1978). To keep costs of construction to a minimum, the fill should be selected from locally available materials. Commonly used fill material and their respective design infiltration rates are presented in Table 6.

Table 6. Commonly Used Fill Materials and Their Design Infiltration Rates.

Fill material	Characteristics	Design Infiltration Rate cm/day	Design Infiltration Rate (gpd/ft^2)
Medium sand[x)]	>25% 0.025-2.0 mm <30-35% 0.05-0.25 mm < 5-10% 0.002-0.05 mm	5	(1.2)
Sandy loam	5-15% clay content	2.5	(0.6)
Sand/Sandy loam mixture	88-93% sand 7-12% finer grained material	5	(1.2)
Bottom ash	---	5	(1.2)

x) Equivalent to 85% by weight between 10 and 60 mesh.

Absorption area. The absorption trench or bed within the mound is sized on bottom area only using the estimated peak daily wastewater flow and the design infiltration rate of the selected fill material.

Mound basal area. The basal area or the fill-natural soil interface of the mound must be sufficiently large to absorb all of the applied wastewater. Once into the topsoil, the liquid can move laterally out beyond the perimeter of the mound until absorbed by the subsoil. On level sites, the entire fill-natural soil interface can be used in determining the necessary area since lateral flow can occur in all directions. On sloping sites, only the area immediately below and downslope from the absorption area is considered. Infiltration rates for the natural soil used in design are presented in Table 7. The soil horizon with the lowest permeability within the upper 60 cm (24 in) should be used in this sizing. Dimensions of other mound components are shown in Figure 4.

Table 7. Infiltration Rates for Determining Mound Basal Area (U.S.EPA, 1980).

Natural Soil Texture	Percolation Rate x)		Infiltration Rate
	min/cm	(min/in)	(gpd/ft^2)
Sand, Sandy loam	0-12	(0-30)	(1.2)
Loams, Silt loams	12-18	(30-45)	(0.75)
Silt loams, Silty clay loams	18-24	(45-60)	(0.5)
Clay loams, Clay	24-48	(60-120)	(0.25)

x) Measured at 30 to 50 cm.

Geometry

The shape of the mound is largely dictated by the characteristics of the site. It is important that the system be laid out such that the water table or zone of saturation does not rise up into the fill material during wastewater application. Therefore, the following must be considered:

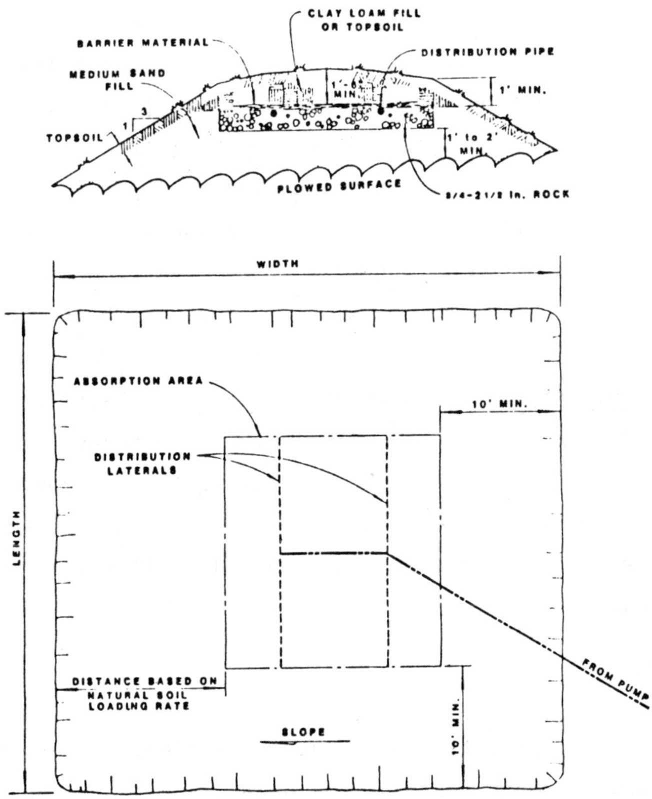

Figure 4. Mound Dimensions.

Soil permeability. If the soil is slowly permeable (slower than 24 to 48 min/cm (60 to 90 min/in), the absorption area within the mound should be a narrow trench 150 cm (5 ft) or less in width oriented with its long axis perpendicular to the natural ground slope. If the percolation rate is faster than 16-24 min/cm (40 to 60 min/in), beds may be used instead of trenches to reduce the mound's length. However, elongated beds with the long axis perpendicular to the natural ground slope are preferred to square beds.

Unsaturated depth. In permeable soils with water tables at 30 to 60 cm (1 to 2 ft), beds no wider than 3 to 4.5 m (10 to 15 ft) should be used within the mound. If the water table is greater than 90 cm (3 ft) below the surface, square beds are acceptable. In slowly permeable soils, perched water table or saturated soil conditions may occur during wet periods. The soil must be carefully examined for any evidence of this.

Layering within the soil profile. The soil profile must be carefully examined for layers which may impede the vertical movement of liquid. If found, long narrow trenches oriented perpendicularly to the natural ground slope rather than beds should be used.

Bedrock or very slowly permeable barriers. Usually, the natural surface topography conforms to the topography of the bedrock surface. If they do not conform, the mound should be oriented relative to the bedrock rather than the ground surface. However, plowing of the natural soil should still follow the surface contours.

Distribution

Although both gravity and pressure distribution networks have been used in mound systems, the pressure networks have been shown to be superior (Converse et al., 1978; Univ. of Wis., 1978). They ensure unsaturated flow is maintained and prevent short circuiting through the fill material by applying the wastewater uniformly over the entire absorption area. It is important that the network is designed such that the manifold drains between dosings. Drainage can be either out the laterals or back into the lift station. The method used depends on the relative elevations of the dosing tank and the distribution laterals. Pressure networks can be designed for simultaneous loading of each absorption area, but dual systems pressurized by duplex pumping units or alternating siphons are preferred to ensure equal divisions of flow. For design of these networks, see Converse (1978); EPA (1980) and Otis (1981).

Construction

Proper construction is extremely important if the mound is to function as designed. Detailed construction procedures are outlined by Converse (1978).

SUMMARY

It is important that onsite systems be sited and designed to prevent surface and groundwater contamination. Therefore, soil and site conditions must be suitable. Where soils are reasonably permeable and have an unsaturated depth of 1.5 m (5 ft) or more, conventional trenches and beds offer a method of wastewater disposal. If these conditions do not exist, then mounds can be often used to overcome slow permeability or shallow soils. Both types of systems must be properly managed after installations, however, to provide a long service life.

REFERENCES

Bendixen, T.W., M. Berk, J.P. Sheehy & S.R. Weibel: 1950, "Studies on household sewage disposal system." Part III. Environmental Health Center, Cincinnati, Ohio.

Bouma, J.: 1975, "Unsaturated flow during soil treatment of septic tank effluent." Journal of the Environmental Engineering Division, ASCE. 101: pp. 967-983.

Converse, J.C.: 1974, "Distribution of domestic waste effluent in soil absorption beds." Transactions of the Amer. Soc. of Agric. Engr. 17: p. 299.

Converse, J.C.: 1978, Design and construction manual for Wisconsin mounds. Small Scale Waste Management Project, University of Wisconsin, Madison.

Converse, J.C., B.L. Carlile & G.W. Peterson: 1978, "Mounds for the treatment and disposal of septic tank effluent." In: Home Sewage Treatment. Proceedings of the second national home sewage treatment symposium. American Society of Agricultural Engineers Publication No. 5-77, St. Joseph, Michigan.

Harkin, J.M., C.J. Fitzgerald, C.P. Duffy & D.G. Kroll: 1979, "Evaluation of mound systems for purification of septic tank effluent." Technical report. Wis. WRC 79-05. Water Resources Center. University of Wisconsin, Madison.

Healy, K.A. & R. Laak: 1973, "Factors affecting the percolation test." Journal Water Pollution Control Federation. 45:1508-19516.

Machmeier, R.E.: 1981,Town and country sewage treatment. Extension Bulletin 304. Agricultural Extension Service. University of Minnesota, St. Paul.

Otis, R.J.: 1981, "Design of pressure distribution networks for septic tank-soil absorption systems." Small Scale Waste Management Project. University of Wisconsin, Madison.

Otis, R.J., J.C. Converse, B.L. Carlile and J.E. Witty: 1978a, "Effluent distribution". In: Proceedings of the second national home sewage treatment symposium, ASAE Publication 5-77. St. Josepth, Michigan.

Otis, R.J., G.D. Plews & D.H. Patterson: 1978b, "Design of conventional soil absorption trenches and beds." In: Proceedings of the second national home sewage treatment symposium. ASAE Publication 5-77. St. Joseph, Michigan.

Thomas, R.E., W.A. Schwartz & T.W. Bendixen: 1966, "Soil chemical changes and infiltration rate reduction under sewage spreading." Soil Science Society of America Proceedings. 30: pp. 641-646.

U.S. Environmental Protection Agency: 1980, "Onsite wastewater treatment and disposal systems" Design manual. Technology Transfer. Cincinnati, Ohio.

U.S. Public Health Service: 1967, "Manual of septic tank practice." Publication No. 526. Washington D.C.

University of Wisconsin: 1978, Management of small waste flows. EPA-600/2-78-173. U.S. Environmental Protection Agency. Cincinnati, Ohio.

Winneberger, J.H., L. Francis, J.A. Klein & P.H. McGanney: 1960, "Biological aspects of failure of septic tank percolation systems". Final report. Sanitary Engineering Research Laboratory. University of California, Berkeley.

PRACTICAL EXPERIENCE WITH ALTERNATIVE SYSTEMS
FOR PROBLEM SOILS IN NORWAY

ALTERNATIVE WASTEWATER SYSTEMS FOR LOCAL INFILTRATION

Nils Skaarer

Landscape Architect, Department of Landscape
Architecture, Agricultural University of Norway,
Aas, Norway

INTRODUCTION

In 1976 a project called "low energy housing" was started near the Agricultural University of Norway at Aas 30 km south of Oslo. The project was carried out in collaboration with the Norwegian Building Research Institute in Oslo and the Foundation of Scientific and Industrial Research in Trondheim. All the 10 testhouses were built in 1978 and were occupied by 35 people.

The project was to be multidisciplinary and included energy saving, small scale wastewater purification plants, biological toilets, recirculation of solid household waste, and use of food-producing plants in the garden.

The ground in the area consisted of a rather dense soil (10-20 percent clay), and about 10 percent of the area had little or no soil.

Three types of local infiltration systems were built and tested, including traditional sand fill built up artificially, shallow trenches supplemented with storage tanks and the mound system. Only the last two will be discussed here.

A. S. Eikum and R. W. Seabloom (eds.), Alternative Wastewater Treatment, 235–241.

All the houses have different types of biological toilets to prevent the outlet of human excretion into the purification system.

BACKGROUND INFORMATION FOR THE ALTERNATIVE WASTEWATER SYSTEM

The presence of rocky terrain in Norway makes sewers very expensive. From a cost standpoint it is important to minimize the need for sewers. Thus on-site disposal of wastewater becomes an attractive alternative.

Very often there is dense or shallow soil with high groundwater. Under these circumstances problems with the treatment of the wastewater frequently arise.

There are two points which are of great importance: the possibility of reducing the effluent volume from each house and the prevention of nutrient and organic materials in the waste. If this can be done, new economical technology can be used to treat gray water.

SHALLOW TRENCHES AND STORAGE

Figure 1 shows some details of this test installation. One family was connected to the purification system. The system was placed into use in the summer of 1978. At first, the family consisted of 3 people and from 1981, 4 persons. The family had a rather low water-consumption, averaging 74 l/p.d in 1980. (Average consumption for all ten families was 83 $\pm$ 20 l/p.d.) The important parts of the system were: septic tank, 30 m^3 storage, automatic pumps, and 30 meters of pipes in shallow trenches.

The system pumped sewage water from April to December, and in the 4 remaining winter months the water was stored in the tanks. The storage tanks were designed to overflow to a sand

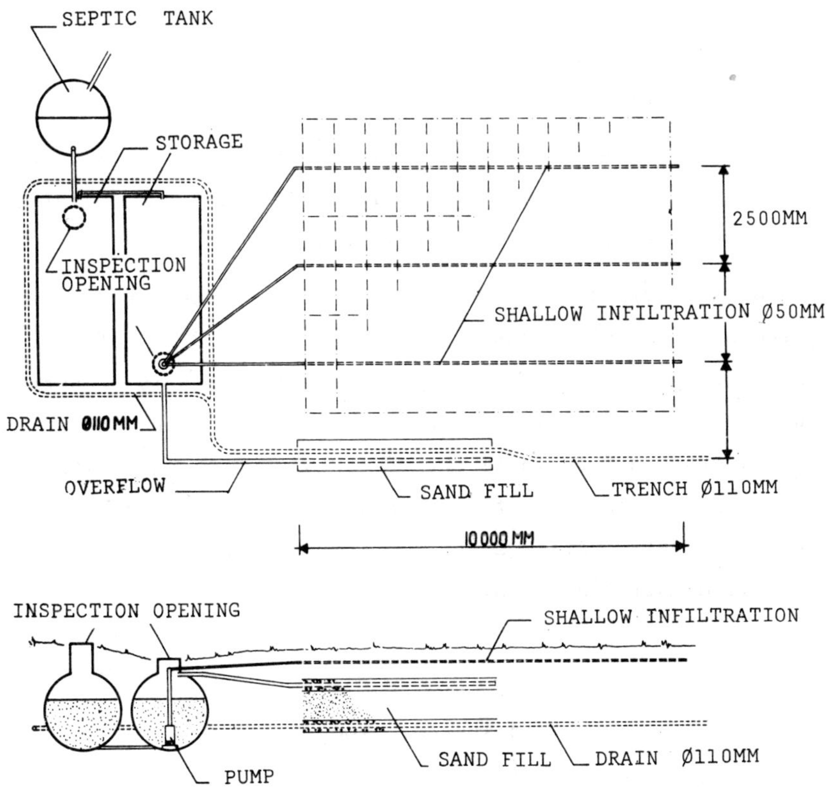

Figure 1. Performing of experiment with storage and infiltration in shallow trenches.

fill, but this has not yet occurred. The system has automatic controls which in the summer activates the pumps four times a day for 20 sec. The outlet-area is 75 m^2 and different species of plants have been tested.

RESULTS

This on-site wastewater system was rather costly. The cost-figure in 1978 was NOK 57 000. The most expensive component was the storage system (NOK 47 000), NOK 5 was about USD 1,00 in 1978 value.

The system functioned very well, and no problems have arisen to date. The shallow pipes have not been blocked by roots or organic components from the sewage water.

Plants such as Rheum cultorum (rhubarb), Ribes nigrum (black currant) and different species of grass (Agrostis stolonifera, creeping bentgrass) have grown very well.

A storage of about 10 000-15 000 liter per person is needed when storage for 4-5 months is involved. About 30 m^2 of infiltration area per person is recommended for the supply of nutrients to the toplayer soil. (It is the same nutrient supply as well nourished grassturf.) There seemed to be no problem in storing sewage water and pumping it into shallow pipes. The plants seemed to grow very well. After 2 years no organic waste or roots were noted in the infiltration pipes.

MOUND SYSTEM

Figure 2 depicts this system. Two families, with a total of 5 people all using biological toilets, were connected to the plant. The mound was put in use in the summer of 1978.

The mound-system consists of a septic tank, pumping tank, two perforated infiltration pipes placed in gravel with a sand fill below (filter area 56 m^2). The mound has sandy soil over and under the sandfill.

RESULTS

The cost of the system was relatively low, about NOK 15 000 for each family (1978). The cost is strongly dependent on transport costs of the mass needed for the mound.

No major problems were noted using this system. The shallow pipe from the tank to the sand fill froze once (it is now insulated). On the northeast corner of the mound water collected at

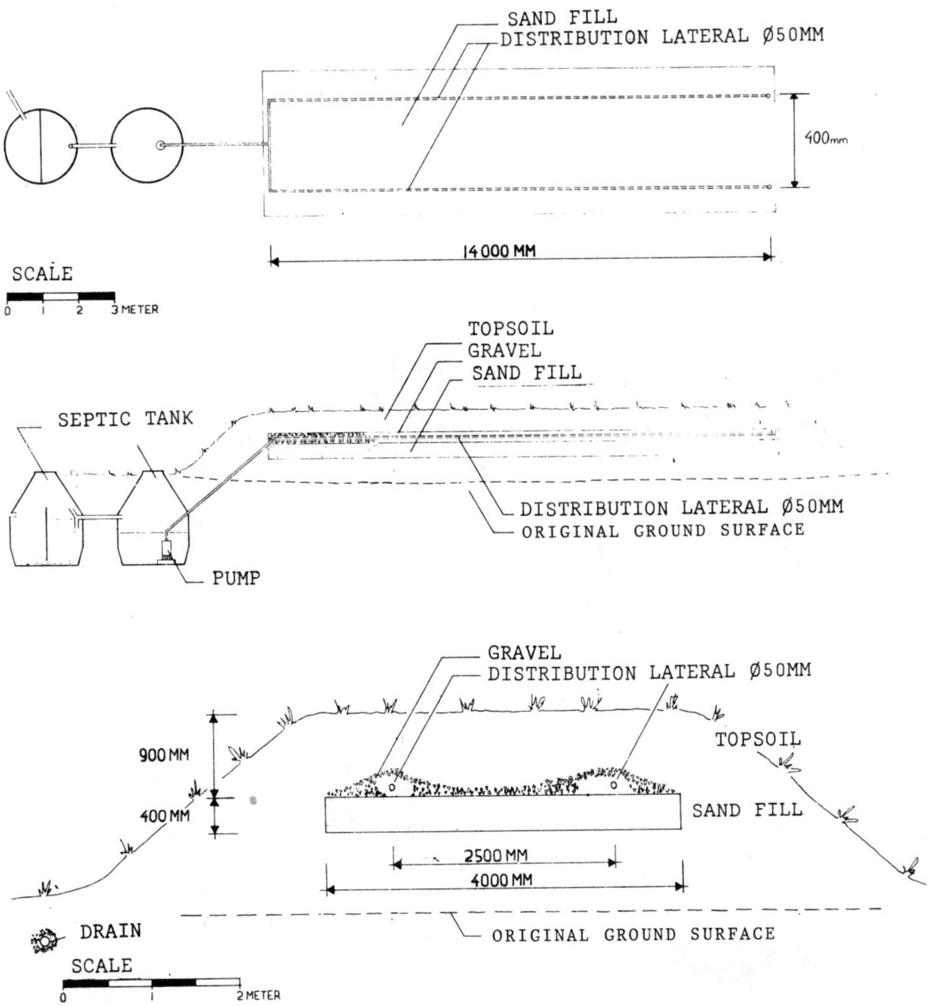

Figure 2. The mound at Danskerud in Aas.

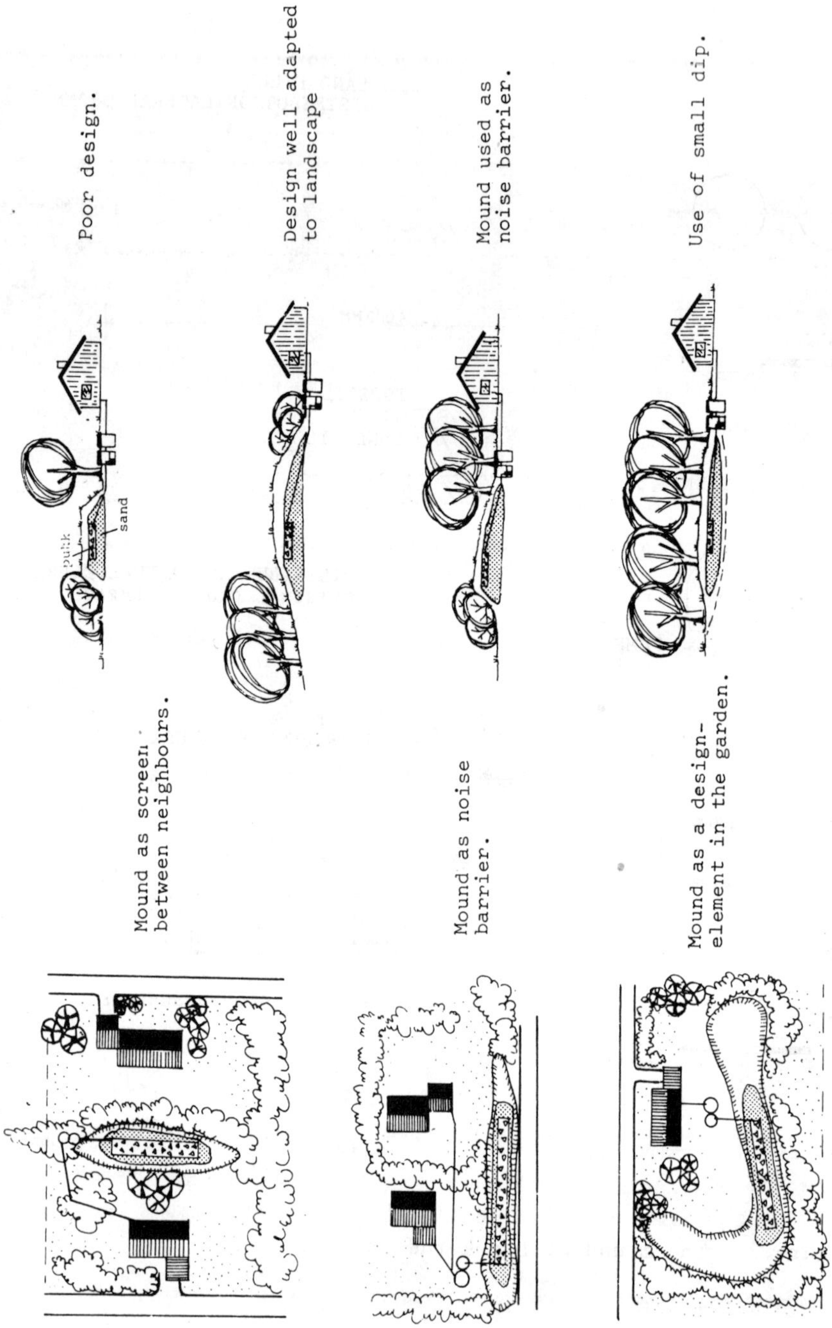

Figure 3. Examples of landscape usage of mounds.

times. There the soil was of a more open structure than elsewhere, and the infiltration rate seemed to decrease after some use. The first year the average infiltration was about 30 cm/day. One year later the infiltration was only 7 cm/day. How low the infiltration will eventually be, is not clear.

The mound was planted with Ribes nigrum (black currant), grass and Trifolium reptans (clover). These species grew very well, but were somewhat surpressed by dry conditions on the south slope.

The system seemed to have been built too high. The soil under the filter had no function. The sand filter should also have covered the base of the mound completely as in the Wisconsin mound.

There has been no use of the mound as a landscape element in this particular experiment. However, there are many possibilities of doing this in an interesting way, as shown in Figure 3.

REFERENCES

Skaarer, N.: 1980, "Rensing av avløpsvann ved infiltrasjon i rotsonen (resorpsjon) kombinert med lager. Undersøkelser på Danskerudfeltet i Ås." (Treatment of wastewater by infiltration in the rootzone (resorption) combined with storage. Survey at Danskerud, Aas.) Report No. 3. Agricultural University of Norway. 23 pp.

Skaarer, N.: 1981, "Resultater fra forsøksfelt i Ås: Vannforbruk i boliger uten vannklosett langt lavere enn ventet." (Results from research-field at Aas: Water consumption in households without water closet far lower than expected.) BYGG No. 5A/Teknisk Ukeblad No. 26A. 2 pp.

Skaarer, N.: 1981, "Bruk av jordhauginfiltrasjon til behandling av avløpsvann. Undersøkelser på Danskerudfeltet i Ås." (Use of mound infiltration for treatment of wastewater. Survey at Danskerud, Aas.) Report No. 8. Agricultural University of Norway. 20 pp.

SESSION VII

NON WATER CARRIAGE SYSTEMS

Chairman: R.W. Seabloom

NORWAY INTRODUCES QUALITY STANDARDS FOR BIOLOGICAL TOILETS

Ove Molland

Senior Research Officer, Section of Soil Pollution Research, The Agricultural Research Council of Norway, 1432 Ås - NLH, Norway

INTRODUCTION

The Department of Microbiology at the Agricultural University of Norway has since 1971 carried out tests of biological toilets for use both in all-year houses and cabins. Most biological toilets on the Scandinavian market have been laboratory tested in order to determine capacity, care, maintenance, etc. The tests have provided guidance for the users of the toilets and stimulated further development (Guttormsen et al., 1975; Guttormsen et al., 1978; Guttormsen & Pedersen, 1978).

In 1977, at the request of the Ministry of Environment, an investigation of biological toilets in practical use was started. The investigation included toilets in all-year houses at the lake of Mjøsa and in cabins in the mountains and at the coast. About 130 toilets have been examined. The project will be finished in 1981. Some preliminary results have been published (Øberg, 1979; Øberg & Molland, 1981).

In 1980 the work with laboratory tests of biological toilets was transferred to the Section of Soil Pollution Research, under the Agricultural Research Council of Norway.

A. S. Eikum and R. W. Seabloom (eds.), Alternative Wastewater Treatment, 245–254.

WHAT IS A BIOLOGICAL TOILET?

A biological toilet is a sewer-less on-site system where feces, urine, toilet paper and organic kitchen wastes are collected, excess water is evaporated and the solid materials decomposed to a soil-like product.

The basic design of the biological toilet is shown in Figure 1a. In order to increase the water evaporation capacity the toilet can be equipped with a heating element and fan. To mix the solid refuse (feces and paper) a mechanical mixer can be installed (Figure 1b).

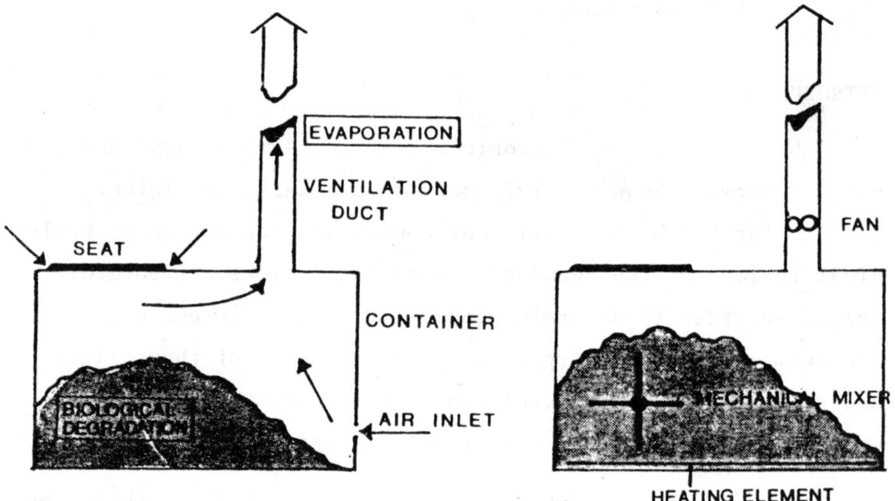

Figure 1. The Basic Design of Biological Toilets.

Biological toilets are divided into two groups according to their size (Figure 2) (Guttormsen, 1980). Small toilets are entirely placed inside the toilet-room. Large biological toilets have the toilet-stool in the toilet-room, while the decomposition tank is either placed in a room below (cellar) or out through the wall. The stool and tank are connected by a pipe.

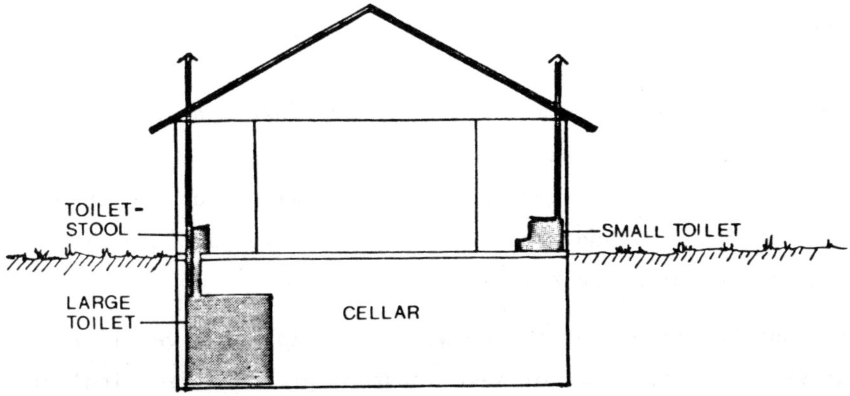

Figure 2. Small and Large Biological Toilets.

All small toilets and the large toilets designed for continuous use are equipped with a fan and heating element in order to increase the capacity. The use and load on toilets will vary in practice. In cabins, the toilets are in periodic use. In Norway they are normally in continuous use for a period of three to four weeks, otherwise just sporadically used during weekends. In all-year houses the toilets are in continuous use.

With the exception of some small toilets that have to be emptied with a frequency of less than three to four weeks, toilets for periodic use most often have larger capacity than toilets for continuous use.

ADVANTAGES OF THE BIOLOGICAL TOILET

The major advantages are:

- they represent a sewer-less, on-site alternative to the water closet.
- they do not use water.
- the waste material is reduced up to 80-90 percent in weight (Guttormsen et al., 1975) and has a dry matter content that makes it easy to handle.

- the comfort for the user is often increased; the toilets can be placed within the house.
- they can be more economical than water closets and water-saving toilets connected to a tank.
- the volume of the wastewater and the total output of dry matter, phosphorus, and, in particular, nitrogen are greatly reduced (Kristiansen & Skaarer, 1979).

About 30 percent of the Norwegian population lives in rural areas where construction of sewer systems presents technical or economical difficulties. Sewer-less toilets, in association with infiltration, resorption or sand filtration for treatment of gray water, are in some areas an essential requirement for house-building.

In practice the alternative sewer-less systems for all-year houses are water-saving toilets with tanks and biological toilets. In cabins the biological toilet is usually the only alternative.

DISADVANTAGES OF THE BIOLOGICAL TOILET

In contrast to water closets, the waste material in a biological toilet is treated on the site; it is a "miniaturized sewage plant". This puts demands both on the construction and use of the toilets.

The most common problems are:
- too low capacity, in particular with regard to water evaporation.
- odors outside with downdraft and inside when fan stops.
- flies.
- poor mixing of solid wastes (paper and feces).
- complicated care and maintenance.
- difficult installation.

FUNCTIONAL TESTS OF BIOLOGICAL TOILETS

Experience with biological toilets in Norway has shown that not all types on the market function satisfactorily, due to poor construction of the toilet, incorrect installation or improper use. The Norwegian State Pollution Control Authority (SFT) has therefore found it necessary to introduce functional tests of the biological toilets. The object is to assure that the units function as intended and to give the consumers better guidance on the selection, installation, and operation of the toilets. The test, which the producers pay for, is voluntary.

The testing procedure has been worked out by a committee with representatives from the public authorities, science, industry and the consumers' organisation (SFT, 1981). Swedish authorities are expected to approve the tests and make them valid also in Sweden.

The tests are administered by the Norwegian State Pollution Control Authority and carried out at the Section of Soil Pollution Research, the Agricultural Research Council of Norway. The administrative process is shown in Figure 3.

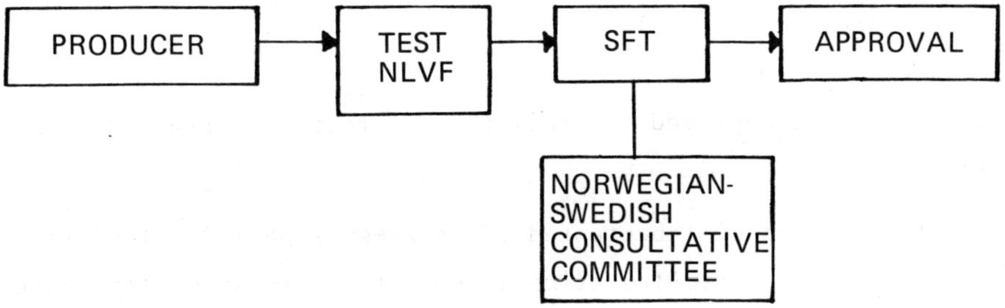

Figure 3. The Administrative Process for Testing the Biological Toilets in Accordance with the Quality Standard. (NLVF: The Agricultural Research Council of Norway; SFT: The Norwegian State Pollution Control Authority.)

After the toilets are eventually approved, each toilet produced will be equipped with a certificate on which is given the following information:

- whether the toilet is constructed for periodic or continuous use.
- capacity.
- energy consumption.

Testing of three small and four large toilets for continuous use started in June 1981 and is expected to end early in 1982.

TESTING PROCEDURE

Toilets for both continuous and periodic use are to be tested. The procedure for testing toilets for continuous use has been established and will now be discussed. The test of toilets for periodic use is not yet definite and will not be dealt with in this paper.

The aim of the test is to investigate:

- instructions for installation and use.
- materials and construction.
- necessary care and maintenance.
- capacity.
- the degraded material.
- odor.

In order to be approved the toilets must fulfill certain requirements.

The toilets will be tested for a 22-week period. They are operated under controlled climatic conditions in specially-built laboratories:

- small toilets at 18-22 ^{o}C and 30-50 percent relative humidity (RH).
- large toilets at 8-12 ^{o}C and 50-60 percent relative humidity (RH).

It is common in Norway to have cellars under the houses, in which the tanks of large toilets can be placed. The climate in the cellars is relatively stable, 10 °C and 50-60 percent RH being representative.

The toilet is loaded with waste material in accordance with the number of persons for which it is designed. The amount of waste used (per person and day) are:

Feces [1,2]	200 g
Toilet paper [3]	10 g
Urine [2,4,5]	1290 g
Total	1500 g

1) as a substitute for raw feces, mechanical/chemical, dewatered (20 percent dry matter (DM)) sewage sludge is used.

2) Gotaas, 1956

3) the average use reported from an investigation of biological toilets at lake Mjøsa (Stein Øberg, personal communication).

4) urea: 30 g; NaCl: 16 g; KH_2PO_4: 4 g; water to 1290 g.

5) Schreiner, 1967.

Toilet paper and feces are added once a day. Urine is automatically added eight times per day. Waste materials are added every day for 22 weeks, except for a period of two weeks.

REQUIREMENTS OF THE TOILETS

In order to be approved the toilet must fulfill certain requirements with regard to construction and function. The main requirements are:

Instructions

Instructions for installation and use shall be included with each toilet. It must clearly be stated how the toilet should be installed, used, and maintained in order to keep it functioning as intended.

Construction

The toilet must be constructed and designed so that no injury to the user can take place. If the toilet-stool and tank are connected by a pipe, the inner diameter of this should not exceed 20 cm in order to prevent small children falling into the tank.

Care and Maintenance

It must be possible to take care of, maintain and empty the toilet in an acceptable way, without having to handle fresh waste.

Capacity and Function

Dry matter content. The decomposed material to be taken out of the toilet must be dry enough so that no drainage of liquid occurs. This is tested by measuring the water retention of the material. The dry matter of the material must equal or exceed the dry matter content found by the water retention test.

Reduction of organic matter. The degree of stabilization of the decomposed material is measured as the reduction in organic matter (loss on ignition) (Schultz, 1962):

$$\text{percent reduction in organic matter} = \frac{A_d - A_R}{Ad-(100-A_R)} \cdot 10^4\%$$

where: A_d = mean loss on ignition of decomposed material (%)

A_R = mean loss on ignition of refuse (%) as calculated from the mass of the various waste materials added and their loss on ignition.

The reduction must at least be 30 percent.

Hygiene. The number of fecal coliform bacteria must be less than 500 per gram of sludge.

Drained liquid. Liquid should normally not be drained from the toilet. If it is, the liquid must meet the following criteria:

- COD less than 2 g per person and day.
- P less than 0.1 g per person and day.
- N less than 0.2 g per person and day.
- less than 500 fecal coliforms per 100 ml.

Odor

There should normally not be any unpleasant odor in the toilet-room.

Materials and Construction

The following requirements are set:

- there must not be any leakage of liquid.
- the toilet stool must not be deformed by a load of 200 kg.
- all parts must resist a blow by a 3 kg iron-ball which with a 75 cm long arm is pulled out at 45° angle and dropped.
- no part of the toilet must ignite when heated with a burning match.
- the toilet must not crack when filled with water and frozen.
- the material in the toilet or mechanical components must not be visibly damaged after the test.

Because biological toilets represent an on-site solution, they have achieved increasing popularity. Today approximately 5000 toilets, divided among some 20 brands, are sold in Norway per year, a figure that is supposed to increase during the coming years. However, in spite of being on the market for years, today's biological toilets must be considered as "the first generation". Undoubtedly, the toilets can be improved both with regard to capacity and efficiency.

REFERENCES

Gotaas, H.B.: 1956, "Composting." WHO, Geneva, 204 p.

Guttormsen, D.M., O. Lind & T.A. Pedersen: 1975, "21 biological toilets. Composting toilets for cabins." (In Norwegian.) Forbrukerrapporten 10, pp. 23-37.

Guttormsen, D.M., O. Lind & T.A. Pedersen: 1978, "Biological toilets." (In Norwegian.) Forbrukerrapporten 3, pp. 18-29.

Guttormsen, D.M., T.A. Pedersen: 1978, "Alternative toilets for cabins and houses." (In Norwegian.) PRA 21, 87 p.

Guttormsen, D.M.: 1980, "Features of different types of "Non-waterborne" toilet systems." Eighth CIB Congress, Oslo, June.

Kristiansen, R. and N. Skaarer: 1979, "Volume and composition of the gray water." (In Norwegian.) VANN 2, 6 p.

Schreiner, A.: 1967, "The anatomy and physiology of man." (In Norwegian.) J.W. Cappelens forlag, pp. 137-139.

Schultz, K.L.: 1962, "Continuous thermophilic composting." Appl. Microbiol. 10, pp. 108-121.

The Norwegian State Pollution Control Authority: 1981, "Quality standards for biological toilets." 18 p.

Øberg, S.: 1979, "Status-report 1. Preliminary results from practical examination of biological toilets for use in all-year houses." (In Norwegian.) Dep. og Microbiology, The Agricultural University of Norway, 32 p.

Øberg, S. & O. Molland: 1981, "Biological toilets. Practical experiences and standards." In: Results from experiments of 1980. (In Norwegian.) Section of Soil Pollution Research. The Agricultural Research Council of Norway. Report No. 2, pp. 67-75.

SESSION VIII

ON-SITE WASTEWATER TREATMENT METHODS

Chairman: H. Ødegaard

PRESENT TECHNOLOGY IN NORTHERN EUROPE ON WASTEWATER TREATMENT PLANTS FOR SMALL FLOWS

Christen Harr

Sanitary Engineer, Trygve Elind A/S,
Engineering and Contracting Company, Oslo, Norway

INTRODUCTORY STATEMENT

This paper gives a survey of wastewater treatment plants for small flows presented on the Norwegian market. These plants have been developed mainly in the northern European countries such as Sweden, Finland, and Switzerland. They are imported as package plants or produced in Norway by a license agreement.

The term wastewater treatment plants for small flows or on-site wastewater treatment plants is used for facilities which serve less than 50 person equivalences.

The State Pollution Control Agency in Norway and the Norwegian State Department of Environment have taken a conservative attitude towards the use of on-site treatment plants. This is due to negative experiences with previous plants in use. For example, the lack of adequate maintenance has resulted in poor performance and many malfunctions. There is no long-term experience in Norway with the available plants on the market. Performance data presented in this paper are based upon figures and numbers from other northern European countries.

A. S. Eikum and R. W. Seabloom (eds.), Alternative Wastewater Treatment, 257–275.

AUTHORITY REGULATIONS

Norway presently has official guidelines for the design of conventional treatment plants, that is primary, biological, biological-chemical, and chemical treatment plants.

In the introductions to these guidelines it states:

> As a rule 50 person equivalences (p.e.) can under satisfactory conditions be set as a lower limit for the use of conventional treatment plants. Satisfactory conditions mean continuously running plants with adequate operation and maintenance done by local water authorities. Strict technical limits for how small a treatment plant can be made are difficult to determine.
> Today there are manufactured biological, chemical and biological-chemical plants that can treat small flows down to 4-6 persons or one household. These plants can with satisfactory operation and maintenance give the necessary performance. However, experiences with such plants have been negative, mainly because of the lack of frequent surveillance.

The State Department of Environment states in a regulation published April 22 1975 that treatment of small flows up to 7 households (25-30 persons) shall be septic tanks followed by infiltration systems. In spite of these regulations there is a need for an alternative to infiltration systems. Consequently, there is a great interest in reliable on-site package plants in Norway today.

GENERAL DESCRIPTION OF PACKAGE PLANTS FOR ON-SITE WASTEWATER TREATMENT

Before each available type on the market is examined in detail, it is appropriate to give some general classification:

- Treatment plants developed especially for on-site wastewater treatment
- Treatment plants developed for large flows and scaled down to serve small flows.

The last type is usually much more complex and requires more intensive care by qualified maintenance personnel.

The first type mentioned is designed with as little mechanical equipment as possible. The design is simple and these plants are meant to function with very little maintenance and operation control. This will be discussed later in the paper, but it is important to note the following:

- Regardless how simple an on-site treatment plant is designed, and no matter how fool-proof the equipment seems to function, malfunction will occur. It is therefore necessary for all on-site treatment plants to require routine operation and maintenance to minimize break-downs.

The following treatment processes are presently available as package plants:

- Septic tanks

  Due to cold climate they rarely give any anaerobic treatment and are looked upon only as settling chambers or primary treatment.

- Biological treatment with aerobic units:

  Activated sludge systems

  Fixed film processes → Biofilter / Biodisc

- Biological-chemical treatment

  Simultaneous precipitation

  Post precipitation

- Chemical treatment (phosphorus removal)

  Precipitation after or in combination with primary treatment.

Figure 1 shows the general process scheme for on-site treatment plants.

The pretreatment and primary sedimentation take place in septic tanks which usually have 3 chambers. The last chamber is very often used as pumping station and balancing tank. The use of a balancing tank is highly recommended. The reduction of peak-flows will improve all sedimentation units based upon continuous flows. A balancing tank will therefore prevent hydraulic

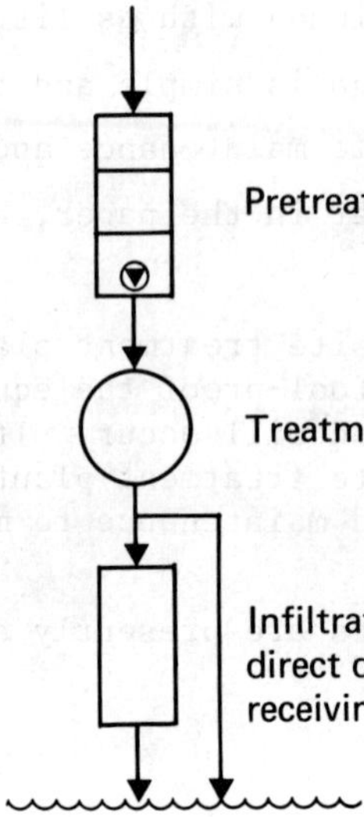

Figure 1. General process scheme for on-site treatment plants.

overloading problems. All package plants on the market today have incorporated balancing tank, or the manufacturer requires some sort of balancing installations if performance guaranties are to be given.

Depending upon local requirements, the effluent from the package plants are usually discharged directly to receiving water or to the infiltration systems.

One of the most important operation and maintenance routines is to assure a satisfactory excess sludge withdrawal, which is usually done by tanker trucks. Figure 2 illustrates sludge production as a function of the treatment process. By using precipitation-chemicals the sludge production increases and in most cases also the operation and maintenance problems.

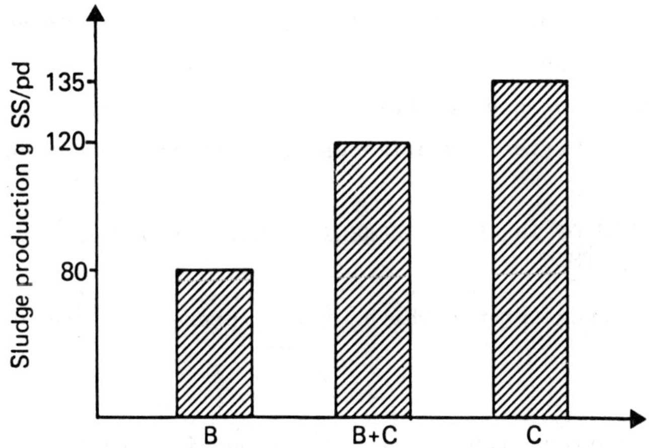

B = Biological treatment including primary treatment

B+C = Biochemical treatment including primary treatment

C = Chemical treatment including primary treatment.

Figure 2. Sludge production as a function of treatment process.

EVALUATION OF PACKAGE PLANTS

The Norwegian Pollution Control Agency established in 1977 a council for evaluation of package plants available on the Norwegian market. The council consists of representatives from:

- The State Pollution Control Agency (Chairman)
- Municipality authorities
- Consultants
- Manufacturers of package plants
- Sewage works personnel.

The council published an evaluation of each plant based upon official design guidelines and operation experiences. Quality and reliability of the construction materials and machinery equipment were also evaluated. Unfortunately, the council did not provide any evaluations of small flow on-site treatment plants because of the previously mentioned regulations. Hitherto there has not been any long term testing of these promising new plants. This needs to be done because it is generally difficult to transfer other countries' experiences to Norwegian conditions.

PRESENTATION OF THE MOST CURRENT PACKAGE PLANTS FOR ON-SITE TREATMENT

The package plants which will be presented, are shown in Table 1.

Table 1. Current package plants for on-site treatment on the Norwegian market (1981).

Commercial name	Treatment Process
MECANA	Biodisc with rotating filter
PARCA NORRAHAMMAR	Biodisc with sedimentation tank
UPO-VESIMIES	Biofilter in direct combination with septic tanks
EMENDO	Biofilter with sedimentation tank
ALCLEAN	Activated sludge
WALLAX	Chemical treatment with sedimentation

Mecana

Mecana package plant is developed and manufactured in Switzerland. The treatment process is shown in Figure 3. From the septic tanks the wastewater is discharged into the Mecana-plant, which consists of a balancing tank, Mecana biodisc and Mecana rotating filter. Sewage is transported from the

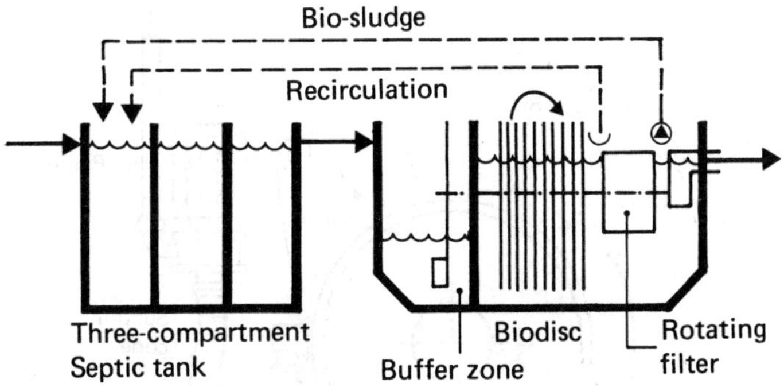

Figure 3. Mecana package plant.

buffer zone to the biodisc by a rotating wheel which has small cups attached to it. Sewage is then lifted at a constant rate "bucket by bucket". The Mecana biodisc consists of PVC discs which are assembled in the form of a spiral and fastened to the center shaft (Figure 4). The number of discs required is dependent on the amount of sewage to be treated by the plant. The disc pack, 1.2 m in diameter, is immersed approximately 2/3 into the sewage.

The biological oxidation process in a biodisc plant is well known and will not be discussed in this paper. Mecana's recommended loading rate on the biodisc is approximately 18,000 mg/l per day, which agrees quite well with Norwegian guidelines. The smallest Mecana plant available has 40 m^2 disc-area and is supposed to provide efficient treatment for 10 persons.

In the Mecana plant the traditional secondary sedimentation tank is replaced by a rotating filter, consisting of a perforated drum, covered by a filter cloth which rotates in the sewage. The sewage must pass through this filter cloth in order to reach the effluent channel (Figure 4). As the filter clogs with sludge, the throughput is reduced. The inflow then becomes greater than

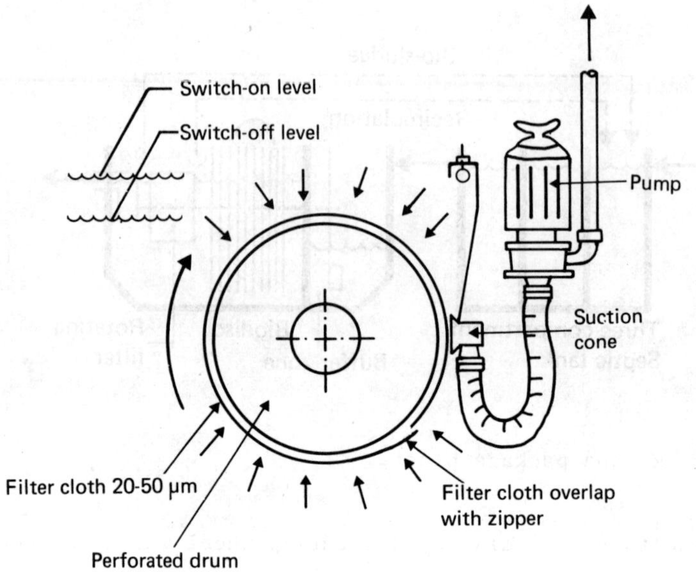

Figure 4. Mecana rotating filter.

the outflow, leading to an increase of the waterlevel inside the tank. At a certain level, a submersible, level-controlled pump starts. The suction end of this pump is connected to a suction manifold placed against the surface of the filter cloth. By this arrangement sludge is removed from the surface of the filter drum within a few revolutions of the filter and is pumped back to the septic tank or to a separate sludge holding tank. This process is automatically repeated at regular intervals. The recommended hydraulic loading rate of the filter is two $l/sec.m^2$.

Operation and performance. Mechanically this plant has been reported as very reliable and needs less surveillance than the more conventional systems. Hydraulic shock loads do not affect the sludge separation process because of the filter construction.

By means of small cups fixed to a rotating wheel at the end of the biodisc unit, an amount of the inlet flow is recirculated back to the septic tank. This will give an O_2 input to the septic tank which possibly may make the sewage more biodegradable when it reaches the biodisc unit.

The filter cloth must be renewed once or twice a year. The performance of the plant is of course dependent on a good filter cloth. There have been some reports on technical problems with the cloth. It is important that the cloth is properly fixed to the drum so that no leakage will occur. It is also important that the cloth has the right density. Routine cloth-changing should be done by service-institutions, either from the manufacturer or the local authorities.

It has been reported that this unit may be able to reduce the effluent BOD to 20 mg/l. Impressive phosphorus removals have also been noted by adding aluminum or iron salts, as shown in Table 2.

Table 2. Phosphorus removal in Mecana package plant with simultaneous precipitation.

Tests	Mol Fe/Mol P	Concentration in effluent	Phosphorus removal in %
1	1.18 : 1	2.4 mg P/l	84.1
2	1.72 : 1	1.05 "	90.9
3	1.40 : 1	1.2 "	98.7

With chemical precipitation the sludge production will increase considerably, and therefore it will be necessary to have more frequent sludge hauling.

Parca Norrahammar

Parca Norrahammar package plant was developed in Sweden. It is designed for capacities down to about 30 person equivalences.

It has been installed at several small institutions, schools, etc., and it is regarded as a compact low cost alternative. The treatment process is illustrated in Figures 5 and 6.

The untreated wastewater is discharged into a primary sedimentation tank which also functions as a buffer tank. An air lift pump transports the primary treated wastewater to a biodisc unit (1.0 m in diameter). The biodiscs, made of corrugated polyester, are inside a closed cylindrical rotating steel tank. The discs are mounted to the tank in such a way that air and water from the air lift pump are forced through the tank from disc to disc.

From the biodisc unit the treated water flows to a secondary sedimentation tank. Sludge from primary and secondary sedimentation tanks is pumped to an aerated sludge-storage tank.

This package plant can also be delivered with equipment for chemical precipitation. The treated wastewater from the biodisc unit is then led to the center of the secondary sedimentation tank where a flocculation unit is installed.

Parca Norrahammar uses 1.1 m^2 disc area/pe. This gives about 0.040 kg BOD_7/m^2.d which is far too much compared to Norwegian guidelines. It is therefore necessary to increase the biorotor area or lower the capacity compared to that given by the manufacturer.

Operation and performance. This plant needs little operation and maintenance. It is a problem, however, to control the biological growth on the discs. Clogging of the biodiscs may occur. It is therefore recommended that a one compartment septic tank be installed in front of the plant, which will then function as a grease trap. There have been some mechanical malfunctions reported with the biorotor bearings, a problem which, according to the manufacturer, has now been solved.

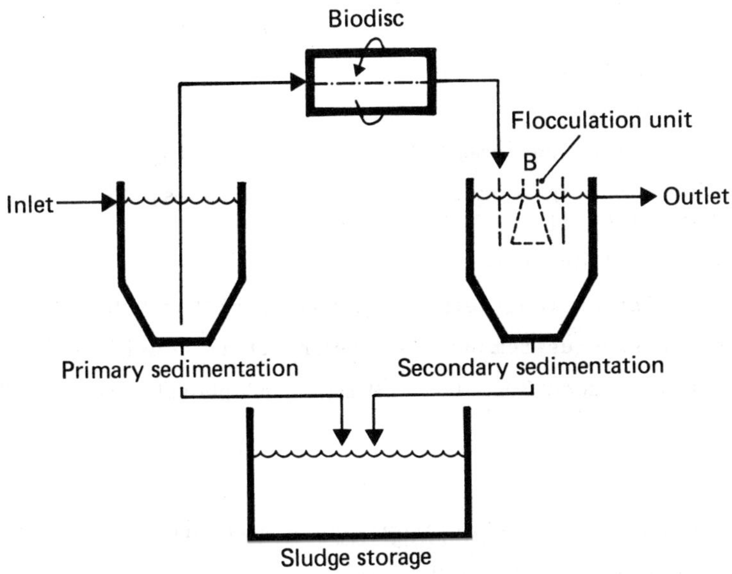

Figure 5. Parca Norrahammar package plant.
Process scheme.

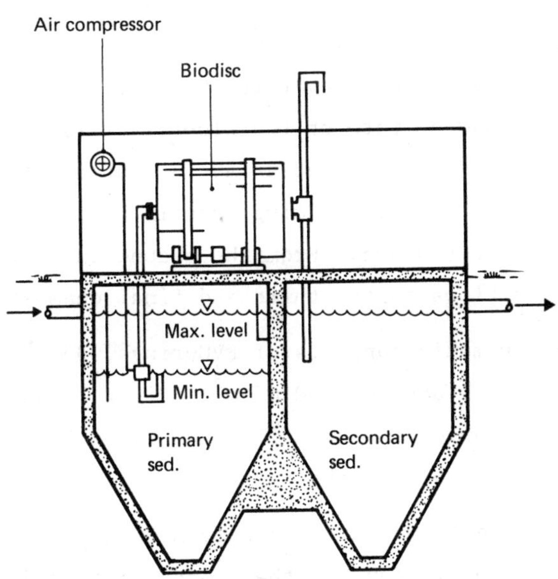

Figure 6. Parca Norrahammar package plant.

It is also necessary to carry out the following maintenance schedule two to three times a week:

- Check air compressor
- Check mechanical equipment for biorotor bearing
- Check for floating scum in sedimentation tanks, hose down sidewalls
- Pump waste solids as required.

No reports exist on long term operating tests for BOD_5, suspended solids, or phosphorus removal. However, it is considered that the plant will reduce BOD_5 to $\leq$ 20 mg/l and phosphorus to 1.0 mg/l.

Upo-Vesimies

Upo-Vesimies package plant was developed and manufactured in Finland. It was developed especially for on-site treatment of wastewater from one to seven households. The treatment process is illustrated in Figure 7.

Untreated wastewater comes into the first compartment of the three-compartment septic tank. From the second compartment wastewater is pumped by a submersible pump to the top plate which distributes the wastewater evenly over the bio-filter surface. The average hydraulic loading rate on the bio-filter is about 3 m/h and the wastewater is recirculated an average of about six times over the filter. The organic loading rate is about 0.7 kg BOD_7/m^3.d filtercolumn. From the second compartment the treated wastewater flows to the third and final chamber.

The bio-filter media consists of PVC-rings, 35 mm in diameter, and a specific surface area of 250 m^2/m^3. Figure 7 shows more in detail the function of the bio-filter unit. A fan at the top of the filter provides air to the filter.

Operation and performance. The only moving parts in this package plant are the air fan and the submersible pump. The pump is connected to an electronic timer, or the operator may

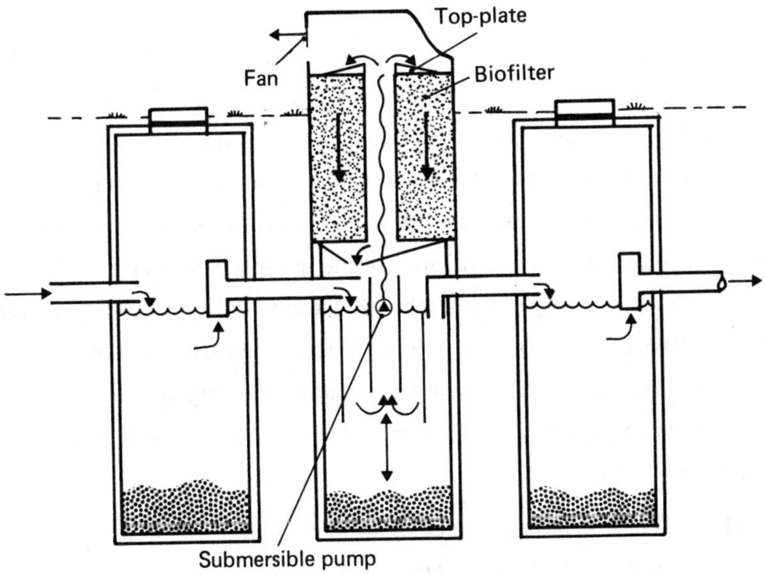

Figure 7. Upo-Vesimies treatment plant.

adjust the pumping cycle. An intermittent pumping assures enough oxygen to the bio-filter treatment process.

The pump must be placed in a position which gives a satisfactory buffer zone in the second septic tank compartment. The septic tanks must be emptied four times a year.

Several problems with this plant have been reported:

- Malfunction of the pump
- Malfunction of the timer
- Clogging of the small diameter pressure pipe from the pump to the top-plate
- The compartments have not been emptied.

Further development of this plant may give an interesting alternative. Although no long term performance data exist for this plant, some 1800 units have already been delivered in Finland, Sweden, Norway, and England.

Emendo

The Emendo package plant is another system based upon the use of bio-filter. It has been developed and manufactured in Sweden. The smallest type has a hydraulic capacity of 2 m^3/h or 50 person equivalences. It is also available with capacities up to 250 persons, and therefore is not actually meant for on-site treatment. A brief description of the plant follows, see Figure 8. From a septic tank the wastewater is pumped through a biofilter. Overflow rate is $\sim$ 4 m/h and organic loading is $\sim$ 1.0 kg BOD_7/m^3.d. This is a high-rate loading and considerably exceeds Norwegian guidelines (comparable values are 2 m/h and 0.6 kg BOD_7/m^3.d).

Biologically treated wastewater is then flocculated before entering the sedimentation unit. Bio-chemical sludge is transported to a separate sludge storage tank. The unit may provide treatment efficiencies of 60 percent BOD_7 removal and 90 percent phosphorus reduction.

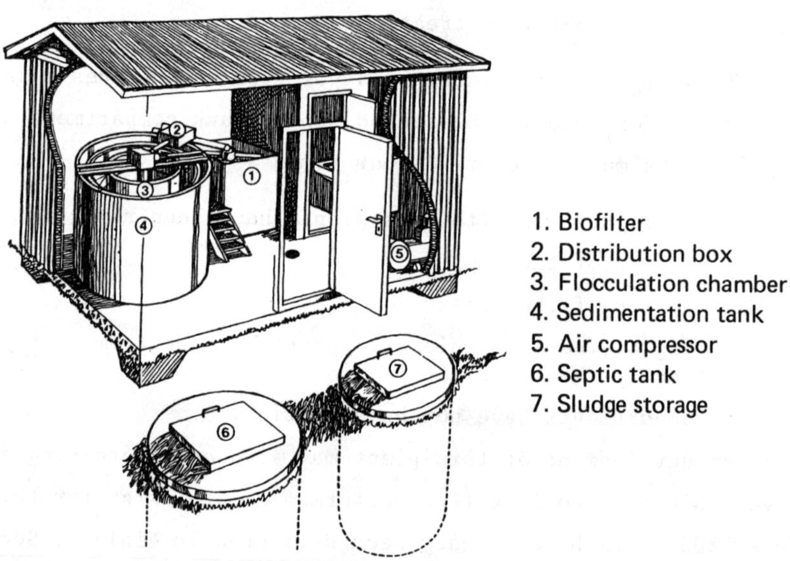

Figure 8. Emendo package plant.

Alclean

This package system is an example of the typical activated sludge plant on the market. Figure 9 is an illustration of the plant. The activated sludge treatment process is well known and will not be discussed in this paper. Instead the operation and maintenance experiences will be evaluated. The activated sludge process can give a very high quality effluent, but needs intensive care. More time must be spent on these units than any of the fixed film-biological processes. It is a delicate process which reacts to changes in flows and inlet waste concentrations. Common operational problems with activated sludge units are listed in Table 3. This table was taken from US EPA design manual: On-site Wastewater Treatment and Disposal Systems (1).

The activated sludge process is also the most energy consuming process due to the use of blowers.

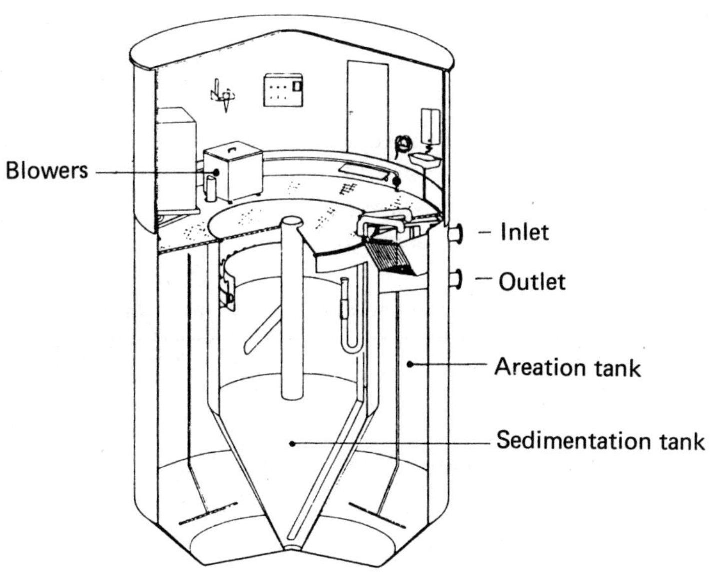

Figure 9. Alclean activated sludge plant.

Table 3. Operational problems - activated sludge package plants. (1).

Observation	Cause	Remedy
Excessive local turbulence in aeration tank	Diffuser plugging Pipe breakage Excessive aeration	Remove and clean Replace as required Throttle blower
White thick billowy foam on aeration tank	Insufficient MLSS	Avoid wasting solids
Thick scummy dark tan foam on aeration tank	High MLSS	Waste solids
Dark brown/black foam and mixed liquor in aeration tank	Anaerobic conditions Aerator failure	Check aeration system, aeration tank D.O.
Billowing sludge washout in clarifier	Hydraulic or solids overload	Waste sludge; check flow unit
	Bulking sludge	See reference (37)
Clumps of rising sludge in clarifier	Denitrification	Increase sludge return rate to decrease sludge retention time in clarifier
	Septic conditions in clarifier	Increase return rate
Fine dispersed floc over weir, turbid effluent	Turbulence in aeration tank	Reduce power input
	Sludge age too high	Waste sludge

Wallax

This is the only chemical treatment plant on the market designed only for on-site purposes. It was developed and manufactured in Sweden. The manufacturer claims that it gives a 70-90 percent phosphorus reduction and 50-70 percent BOD -reduction. Figure 10 gives an illustration of the process. Untreated sewage is fed into the outer tank where primary settling takes

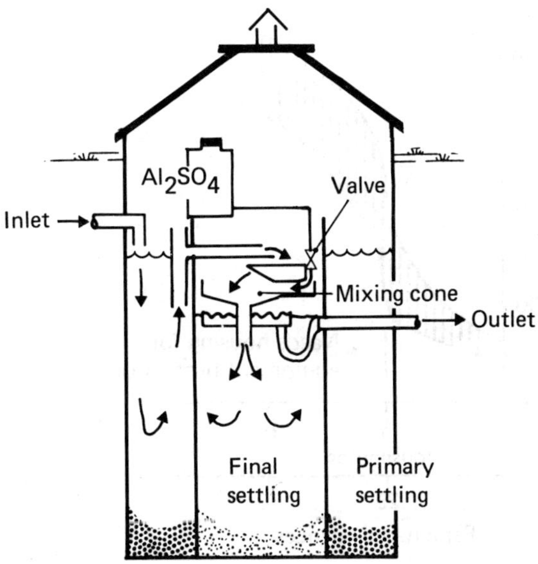

Figure 10. Wallax chemical treatment plant.

place. From here wastewater is discharged through a box which is emptied automatically by means of the balancing weight principle. This box also activates a valve, and aluminumsulfate is dosed and mixed with the wastewater.

Chemical sludge is then settled out in the inner tank. The effluent from the plant is then transported to a sub-surface infiltration system.

This plant is available for one to five households. Sludge must be taken out 4 times a year. No long term performance data are reported. One significant advantage of this plant is that it requires no electricity.

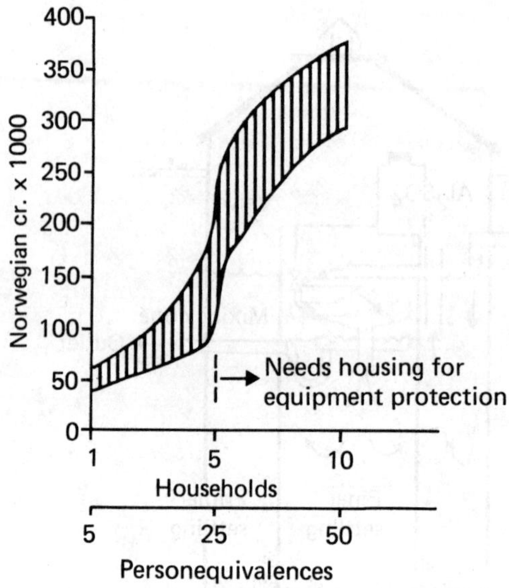

Figure 11. Installation costs for available on-site treatment plants.

COSTS

Figure 11 gives an indication of the installation costs for an on-site treatment plant. The more complex types for several households need protection from the weather. The costs include all construction work necessary for a ready to use installation including pretreatment in a three-compartment septic tank and the biological or chemical treatment unit. The cost is given as a band because the costs will vary with the treatment process. Chemical treatment gives you the lowest costs, biological-chemical treatment the highest costs.

CONCLUSION

The treatment plants discussed in this paper are the most common types and are representative of the types available on the market today (1981). New types from other countries and/or other manufacturers are generally presented every year. Because of restrictive regulations not many are in use in Norway at the present time. The use of such package plants depends on the future policy of the Department of Environment. In the opinion of the author the technology is available presently, and a number of the existing plants on the market can perform reliably. However, it is suggested that every on-site treatment installation be forced to have routine surveillance done by the municipalities. Most of these package plants will be privately owned, but the routine maintenance must be done by skilled personnel.

Special guidelines for the design of on-site treatment plants should be developed. The water authorities should then accumulate long term performance and mechanical reliability data for plants following these guidelines.

REFERENCES

1. "On-site Wastewater Treatment and Disposal Systems, Design Manual". United States Environmental Protection Agency, Office of Water Program Operations, Washington, D.C. 20460, Oct. 1980.

TREATMENT OF RESIDENTIAL GRAYWATER WITH INTERMITTENT SAND FILTRATION

W.C. Boyle, Professor
R.L. Siegrist, Research Specialist
Chin C. Saw, Research Assistant

Department of Civil & Environmental Engineering,
University of Wisconsin, Madison, Wisconsin, 53706, USA

INTRODUCTION

As part of the overall strategy to seek alternative technologies for the treatment and disposal of wastewaters onsite, methods of in-house wastewater modification have been considered. These modifications have been developed as a part of three interrelated strategies; namely, water conservation and wastewater flow reduction, pollutant mass reduction, and onsite containment for subsequent off site disposal. A general schematic of these strategies is depicted in Figure 1.

The chief purpose behind in-house waste modification is to provide a more reliable, manageable, and cost effective solution to the treatment and/or disposal of wastewater onsite. It has been speculated that many of these in-house strategies will achieve this goal, but such prognostications have been based upon literature reviews, speculative calculations, and projected performance of the fixture or treatment element. Few full-scale total system studies have been performed with the proper controls to ensure that the methodology will meet the criteria of reliability, manageability, and cost effectiveness.

A. S. Eikum and R. W. Seabloom (eds.), Alternative Wastewater Treatment, 277–300.

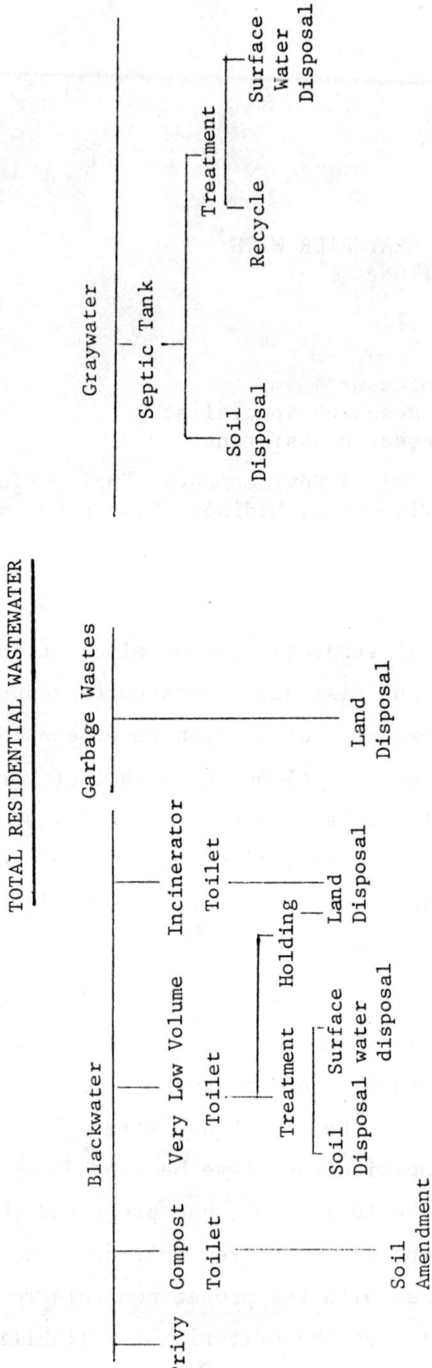

Figure 1. Management Strategies for Segregated Wastes.

One of the most popular segregation systems involves the separation of toilet wastes, garbage wastes, and the remaining washwaters, called gray water. Typical characteristics of these waste fractions are listed in Table 1. It is clear, from examination of this table, that none of the waste fractions are of acceptable quality for surface discharge or direct reuse without some treatment. If such a system is to be adopted, all three waste elements must be treated and disposed of in a manner which is reliable, manageable, without serious environmental impact and cost effective.

Table 1. Pollutant Contributions of Major Residential Wastewater Fractions[a] (gm/cap/day).

Parameter	Garbage Disposal	Toilet	Basins, Sinks Appliances	Approximate Total
BOD_5	18.0 10.9 - 30.9	16.7 6.9 - 23.6	28.5 24.5 - 38.8	63.2
Suspended Solids	26.5 15.8 - 43.6	27.0 12.5 - 36.5	17.2 10.8 - 22.6	70.7
Nitrogen	0.6 0.2 - 0.9	8.7 4.1 - 16.8	1.9 1.1 - 2.0	11.2
Phosphorus	0.1 0.1 - 0.1	1.2 0.6 - 1.6	2.8 2.2 - 3.4	4.0
Flow				
(L/use)	7.6 7.6 - 7.9	16.2 15.0 - 19.0	-	-
(use/Cap/D)	0.6 0.4 - 0.7	3.5 2.3 - 4.1	-	-
(L/Cap/D)	4.5 3.0 - 5.7	6.2 34.8 - 75.6	109.6 81.6 - 131.5	175

(a) Means and ranges of results reported in (1) (2) (3) (4) (5)

The purpose of the research program reported herein was to provide additional information on one key element of this waste modification scheme - the treatment of the graywater fraction. A summary of the results of this study are reported herein. Details of the research program may be found elsewhere (6) (7).

METHODS AND MATERIALS

The field research reported herein was conducted at two sites on the Arlington Research Farm - University of Wisconsin, Madison. The objectives of the study were to

(a) characterize residential gray water at these two sites,

(b) evaluate the effect of filter sand size and gray water application rate on the performance of an intermittent sand filter, and

(c) evaluate the performance of a septic tank, intermittent sand filter as a means of reclaiming residential gray water for effluent disposal or reuse.

Experimental Design

The study was conducted in several phases as indicated in the objectives of the program: graywater characterization; septic tank-intermittent sand filter study; and factorial analysis of sand size and filter loading.

Graywater characterization consisted of sampling graywaters from two homes. In addition, selected graywater samples were analyzed to evaluate settling properties and particle size characteristics.

The septic tank-intermittent sand filter studies were conducted at one home employing a pilot sized septic tank and sand filter system. The sand filter was operated under two loading conditions during the course of this 14 month study.

A replicated 2^2 factorial design was used in a third phase of this study in order to evaluate the effect of sand size and graywater filter loading on intermittent sand filter performance. Sand columns installed in the field at one home were employed in this third phase which was conducted over an 8 month period. The experimental design is presented in Table 2.

Table 2. Factorial Design - Sand Column Study.

Column Number	Hydraulic Loading Rate	Media Size
1, 2	High Rate 33 cm/d	Coarse Sand E_{10} - 1.02 mm U - 1.44
3. 4	High Rate 33 cm/d	Medium Sand E_{10} - 0.17 mm U - 2.82
5. 6	Low Rate 13 cm/d	Medium Sand E_{10} - 0.17 mm U - 2.82
7, 8	Low Rate	Coarse Sand E_{10} - 1.02 mm U - 1.44

Residence Characteristics

The graywater study facilities were installed at rural residences with characteristics as detailed in Table 3. To facilitate the investigation, the existing plumbing systems at both dwellings were altered so that the graywater from all sinks, bathtub/shower, and clotheswasher were collected and transported separately from the home. To eliminate any contamination of the graywater system, a new collection network was installed for the graywater with toilet waste and water softener regeneration waste transported in the existing network.

Table 3. Residence Characteristics.

	Home P	Home S
Composition of family		
Number of people	4	3
Adults [age,sex,occupation]	2 [29,male,poultry worker; 24,female,housewife]	2 [30,male,herdsman; 33,female,housewife]
Children	2 [5,female,child; 3,female,child]	1 [1,female,infant]
Type of building/construction	Existing, single story	Existing, single story
Number of bedrooms	3	3
Water using fixtures		
Number of bathrooms	1	1
Number of toilets	1	1
* Number of bathtubs with shower	1	1
* Automatic clotheswasher	yes	yes
* Automatic dishwasher	no	yes
* Bathroom sink	yes	yes
* Kitchen sink	yes	yes
* Laundry sink	yes	yes
Garbage disposal	no	no
Others		
Water softener	Manual	Manual
Water supply	Private well	Private well
Waste disposal	SSAF	SSAF

* Sources of greywater production used in this study.

SSAF denotes septic tank-soil absorption drainfield.

Sampling Equipment

At each site, a buried sampling well for raw graywater was installed outside the residence. The raw graywater sewer line was directed into and through this well to the septic tank. Inside the well, no-hub couplings were utilized so that the graywater pipeline could be diverted, enabling collection of flow composted samples of the graywater. During sampling, graywater was subsequently pumped from the sampling well to the septic tank.

Quantitative characterization of the raw graywater was accomplished using detailed water use monitoring equipment described elsewhere (6). All 24 hour composited samples were analyzed in accordance with Standard Methods.

Unit Treatment Processes

All treatment studies were conducted at Home P and were constructed outside of the residence (Figure 2). A brief description of each process follows. Detailed descriptions of these processes can be found elsewhere (6) (7).

Septic tank. The septic tank utilized was one of the smallest commercially available units in the area. It was a precast concrete tank, approximately 1.7 m in diameter with a liquid depth of 0.9 m and a total liquid volume of 1985 liters. The inlet and outlets from the tank were baffled with semi-circular fiberglass baffles attached to the tank wall around the inlet and outlet, respectively.

Sand filter. The characteristics of the sand filter are depicted in Figure 3. In selecting the media to be used in the sand filter, achieving a relatively high quality effluent while obtaining reasonably long filter runs ($\geq$ 1 yr) was a primary consideration. Further, high loading rates were desirable to minimize the filter surface area required. Several previous investigations with total household wastewater and domestic wastewater

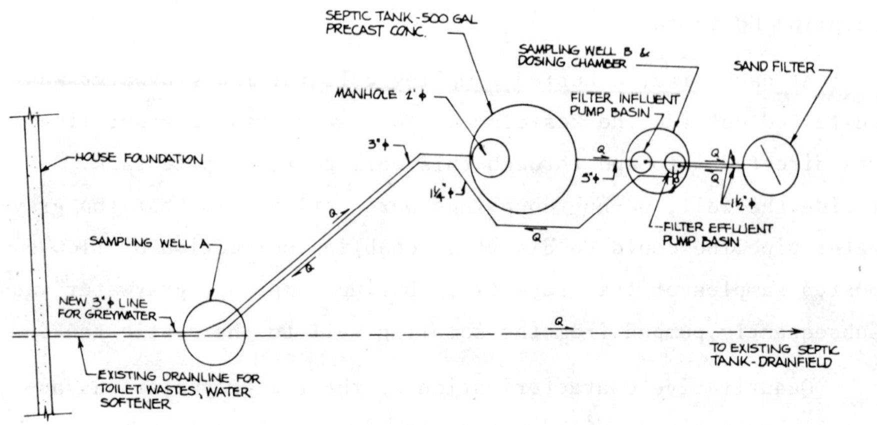

Figure 2. Graywater Treatment Facility

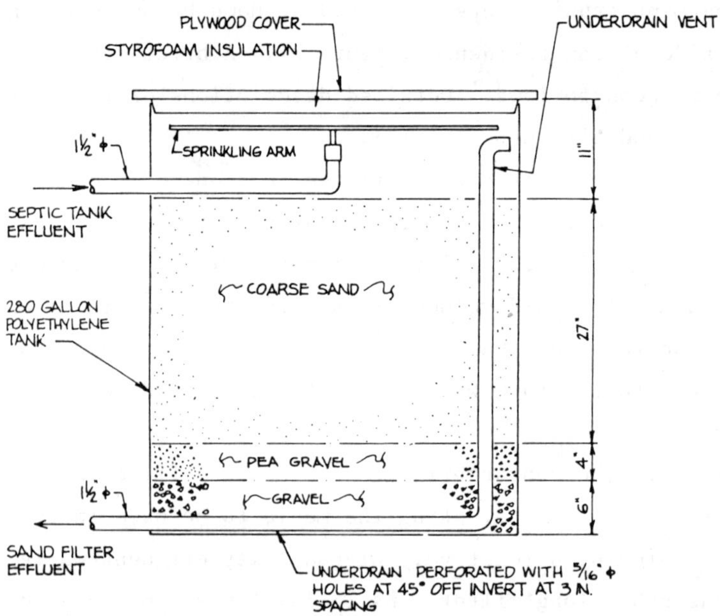

Figure 3. Sand Filter Profile.

had shown that a filter possessing these characteristics was possible (Hines and Favreau (8), Furman et al. (9)). The media finally selected was a coarse filter sand with effective size of 1.37 mm and uniformity coefficient of 1.30. The initial saturated hydraulic conductivity of this sand was 520 m/d.

In sizing the filter, the graywater flow from the study residence was projected to be in the range of 283 l/d based on the results of previous characterization studies and several water meter readings taken at the site. An average daily loading rate of approximately 33 cm/d was selected requiring a filter surface of 0.9 m^2. The filter was 1.2 m deep and 1.07 m in diameter. A section of 3.8 cm diameter plastic pipe perforated with 7.9 mm diameter holes served as the underdrain. Approximately 15 cm of 2.5 cm washed hardrock overlain by 10 cm of 0.6 cm pea gravel served as the support media for 68 cm of filter sand. Uniform distribution of graywater over the sand surface was supplied by a rotating aluminum sprinkler arm 4.7 mm in diameter and perforated with 3 mm holes.

The intermittent sand filter was operated under two different conditions during the test period. In the first test phase, the daily loading was designed to be applied at intermittent doses of 5 cm, with the actual dosing schedule and daily loading determined by the graywater generation pattern in the home. During the second test phase, the daily loading was designed to be applied in hourly doses of approximately 1.4 cm each. A hich-level switch was used to activate extra doses of 4 cm each during periods of high flow.

Sand columns. Eight sand columns were used in a factorial design to evaluate the effect of filter grain size and hydraulic loading on filter performance. These columns were located in the sampling well and dosing chamber following the septic tank (Figure 2). The eight cast acrylic columns were 9.5 cm in diameter and contained approximately 9 cm of 2.5 cm gravel, 6 cm of 0.6 cm pea gravel, and 68 cm of filter sand. Two sand sizes were employed; a coarse sand with effective size of 1.02 mm and uniformity coefficient of 1.44 and a medium sand with effective size of 0.17 mm with a uniformity coefficient of 2.82. Each column was provided with 11 3-mm diameter underdrain vents located approximately 2.5 cm above the bottom.

Graywater from the septic tank was dosed to the surface of each column once per hour at 0.5 or 1.4 cm depending on design filter loading. Extra doses were provided as required by higher than average flows from the dwelling.

RESULTS

Graywater Characteristics

Flow composited samples of raw graywater were collected periodically during the research program. The daily total raw graywater results for both homes P and S are summarized in Table 4. The measured daily per capita generation of graywater in home P was found to be 80.5 l/cap/day based on 121 days of continuous monitoring. No data on flow were available at home S since problems in measurement occurred. A comparison of these results with those found at two other rural households in Wisconsin, a summary of graywater data analyses by EPA (10) and typical residential wastewater are also shown in Table 4.

Analysis of the data presented in this table reveals that including kitchen sink wastes increases BOD_5, suspended solids, and nitrogen levels in the graywater. Depending on the treatment and disposal or reuse scheme considered, it may be advisable to exclude the kitchen sink component from the graywater flow.

Table 4. Summary of Graywater Quality.

Parameter	Units	Graywater: This study, Residence S	Graywater: This study, Residence P	Graywater: This study, Residence NE	Graywater: (6), Residence VAR	Graywater: U.S. EPA (10)	Total residential wastewater U.S. EPA (10)
BOD_5	mg/liter	271	291	125	147	260	200-290
COD	mg/liter	600	622	242	276	---	680-730
TS	mg/liter	925	686	794	810	---	680-1000
TVS	mg/liter	289	274	128	179	---	380-500
TSS	mg/liter	139	136	36	92	160	200-290
TVSS	mg/liter	122	90	33	38	---	150-240
TKN	mg-N/liter	17.4	18.4	5.8	5.7	17	35-100
NH_4-N	mg-N/liter	1.6	4.5	0.6	1.2	---	6-18
NO_3-N	mg-N/liter	0.1	0.6	0.5	0.4	---	< 1
TP	mg-P/liter	11.9	4.8	1.0	0.3	26	18-29
Alkalinity	mg$CaCO_3$/liter	378	382	---	---	---	---
Turbidity	NTU	67	58	42	---	---	---
pH	---	-	7.3-8.2	7.1-8.7	7.2-7.7	---	---
Total coliforms*	log no./liter	9.9	7.90	---	---	---	10-12
Fecal coliforms*	log no./liter	7.8	7.28	---	---	---	8-10
Graywater event**	--	KS,BS B/S, CW, LS		B/S, CW, WS	BS, B/S CW, LS, WS	Basins, sinks appliances	Basins, sinks appliances plus toilets

* Based upon log-normalized data.

**Events included in total stream sampled: KS=kitchen sink; BS=bathroom sink; B/S=bath or shower; CW=clotheswasher; LS=laundry sink; WS=water softener.

The results of this study generally agree with the limited data base available regarding the characteristics of the total residential graywater stream. When compared to total residential wastewater, graywater comprised of all household basins, sinks, and appliances possesses lower concentrations of most conventional pollutants with the exception of BOD_5. If the kitchen sink wastes are excluded, the pollutant concentrations in the graywater are significantly lower than in the total residential waste stream.

A cursory study was also conducted to determine the settling properties and particle size distributions of particles in graywaters. Results of this brief study appear in Table 5. It is apparent that for these two homes, little settleable or floatable material was present in the graywater. Particle size estimates were determined by selective filtration on a variety of membrane and glass fiber filters. It should be noted that this measure was highly subjective and no control of all of the variables that affect filtration was achieved. Yet, this practical measure provided some insight into the applicability of filtration processes on graywater.

Table 5. Size and Settling Properties of Graywater.

		Home P			Home S		
Parameter	Unit	No of data	Mean	Range low - high	No of data	Mean	Range low - high
Particle size Distribution	percent	4			2		
(20-25)	micron		33.25	31.8 - 36.0		26.6	15.5 - 37.7
5 to (20-25)	micron		7.80	7.0 - 8.5		13.8	6.1 - 21.5
0.8 to 5	micron		1.65	1.0 - 2.9		3.35	2.2 - 4.5
0.45 to 0.8	micron		6.20	5.2 - 7.2		6.25	1.7 - 10.8
0.45	micron		51.10	49.3 - 55.0		50.00	-
Settleability	ml/l	4			2		
after 15 min			1.63	1.0 - 2.3		0.20	-
after 30 min			1.95	1.1 - 3.0		0.45	0.4 - 0.5
after 45 min			2.05	1.3 - 3.0		0.55	0.5 - 0.6
after 60 min			2.13	1.4 - 3.0		0.60	0.5 - 0.7
after 90 min			2.15	1.5 - 3.0		0.65	0.5 - 0.8
after 120 min			2.18	1.5 - 3.0		0.75	0.6 - 0.9
Percent settled After 60 min	percent	3	10.9	9.4 - 13.6	2	5.2	3.6 - 6.7

Septic Tank Performance

The average daily flow processed through the septic tank during the 14 month study period was approximately 321 l/d. The average detention time in the tank was 6.2 days, based upon its liquid volume of 1984 liters. During periods of high flow from the home (e.g., on laundry days), the detention time in the tank was substantially reduced.

The maintenance requirements of septic tanks receiving household wastewater are minimal, limited to pumping of the tank contents every few years. Inspection of the septic tank involved in this study after nearly 7 months of operation revealed that minimal amounts of sludge and scum were present. These observations agree with those of Brandes who found that sludge and scum accumulation in a graywater septic tank necessitated pumping approximately every 20 years.

The effluent quality produced by the septic tank is delineated in Table 6. Comparison with results by Brandes (11) and Kristiansen (12) are given also. The septic tank effectively reduced the suspended solids in graywater (70 percent) and appeared to result in ammonification of the organic nitrogen. Even with a residence time of 6 days, little effect was noted on other pollutants, including total and fecal coliforms.

Sand Filtration

<u>Field study - phase I</u>. During the first phase of operation, the sand filter was dosed intermittently with septic tank effluent beginning on November 27, 1979. Application was continued until the hydraulic conductivity decreased to the point where it was continuously ponded between doses. This occurred after 119 days of operation and after loading to the filter totalled approximately 3700 liters, applied in a total of 784 doses. Application to the filter during this period averaged 6.6 doses per day

Table 6. Septic Tank Treatment of Graywaters.

Parameter	Unit	This study	Brandes (11)	Kristiansen** (12)	Septic tank effluent with total residential waste (10)
BOD_5	mg/liter	216	149	107-160	138
COD	mg/liter	502	366	307-370	327
TS	mg/liter	659	528	432-549	---
TVS	mg/liter	279	---	---	---
TSS	mg/liter	52	162	28-43	49
TVSS	mg/liter	37	---	---	35
TKN	mg-N/liter	18.6	11.3	13-26.2	45
NH_4-N	mg-N/liter	10.3	1.7	5.9-17.9	31
NO_3-N	mg-N/liter	0.2	0.2	< 0.05	0.4
TP	mg-P/liter	3.2	1.4	---	13
Alkalinity	mg$CaCO_3$/liter	342	148	---	---
Turbidity	NTU	52	---	---	---
pH	---	6.8-7.6	6.8	---	---
Total coliform	log no./liter	7.96	8.38	---	---
Fecal coliform	log no./liter	7.09	7.15	5.27-6.70	6.7

**Multi-chambered tanks. BOD_7 value shown.

at 5.7 cm/dose, yielding a total daily loading of 38 cm/day. Actual individual filter dosings and daily loadings depended on the graywater generation patterns in the home. For example, for those days where daily flow measurements were available, the daily loading ranged from 7.5 to 105 cm/d.

Visual observation of the sand filter surface with time during the first filter run indicated that the hydraulic conductivity gradually decreased. Initially, no ponding occurred on the sand surface. After 63 days of operation, however, surface ponding occurred until cessation of a dose and the time required for 50 percent of the applied dose volume to outflow from the filter increased to 17 min. After approximately 98 days of operation, the filter surface remained ponded for extended periods and by 119 days the surface was continuously ponded. At that point, loading to the filter was terminated and the filter was allowed to drain. Twenty-four hours later a dose was applied to the filter. After 87 min. the filter surface was still ponded and only 10 percent of the applied dose volume had flowed out to the filter bed.

In an attempt to rejuvenate the filter, the sand media was raked to a depth of approximately 10 to 15 cm. A very thin, grayish-colored mat was present on several locations of the filter surface and a black discoloration of the sand extended 10 to 15 cm into the filter bed. After raking, the filter bed surface was smoothed out and a second filter run was initiated.

The initial hydraulic conductivity of the filter after raking was similar to that of clean sand, but after only 6 days, a significant decrease was evident. After 27 days of operation, the filter surface was continuously ponded and the filter run was terminated. The filter was allowed to drain naturally for 48 hr. and, then, a dose was applied to enable observation of the outflow rate. After 42 min., 10 percent of the applied dose volume had passed through the filter and, after 59 min. the filter surface was still ponded slightly. The filter was again rejuvenated by raking and, then, reloaded. A third run of approximately 28 days was achieved prior to continuous surface ponding and failure.

During the initial period of this study, the influent wastewater temperature was relatively constant at 8 to 10 ^{o}C, while the air temperature over the filter fluctuated between 2 and 10 ^{o}C. Supplementary heat was added subsequently above the filter, stabilizing the air temperature to about 10 ^{o}C.

The effluent quality achieved with this sand filter operation is presented in Table 7. Removals were very poor and virtually no nitrification occurred.

Table 7. Reduction of Mean Pollutant Concentrations in Graywater by Sand Filtration - Phase I*.

Parameter	Unit	Septic tank effluent	Sand filter effluent**	% reduction
BOD_5	mg/liter	215	83	61
COD	mg/liter	473	232	51
TSS	mg/liter	41	45	0
TVSS	mg/liter	22	4	82
TN	mg-N/liter	22.0	14.8	33
Turbidity	NTU	52	20	62
Total coliform †	log no./liter	8.08	7.25	85
Fecal coliform †	log no./liter	7.27	6.55	81
Period		11-27-79 to 5-28-80	11-27-79 to 3-3-80	---

*Based upon 24-hr flow composited samples.
**Effluent produced by intermittent filtration through 27 in. of coarse sand (E.S. = 1.37 mm, U.C. = 1.30) at a mean daily loading of 8.7 gpd/ft^2.
†Log-normalized data.

These limited removals were likely due to several causes. The filter was started up in cold weather providing little opportunity for proper bed maturation. The coarse media employed and the high application rates also may have contributed to this situation. It was surprising that the filter run was only 119 days using this coarse sand.

Field study - phase II. To enhance the poor performance exhibited by the filter in phase I, a new mode of operation was selected as described earlier. Prior to initiation of this phase, the top 25 to 30 cm of the filter bed was removed and about 11 liters of 10 percent H_2O_2 were dosed through the filter bed. After flushing with tap water, clean sand was added.

Operation of the filter began on July 15, 1980. Application to the filter continued until the hydraulic conductivity decreased to the point where continuous ponding occurred. This occurred after 149 days of operation during which time the loading to the filter totalled approximately 49,000 liters applied in a total of 3,348 hourly doses (11 liters/dose) and 326 high-level doses (33 liters/dose). The total daily loading to the filter during this period averaged approximately 39 cm/day. This loading was applied in an average of approximately 22 hourly doses/day and two high-level doses/day. Actual individual filter doses and daily loadings depended on the home graywater generation patterns. During the 90 days where detailed data were available, filter doses ranged from five to 25 hourly doses/day and from 0 to 13 high-level doses/day. The daily filter loading ranged from 7 to 85 cm/d for the data period. During a high flow period, such as on a laundry day, up to five high-level doses were recorded within a 20 min. time period.

Visual observation of the sand surface during this filter run revealed a slowly decreasing hydraulic conductivity. After 86 days of operation, surface ponding occurred for only 0.2 min. after cessation of a high-level dose and 50 percent of the applied dose passed through the filter in 11.8 min. Subsequently, the filter conductivity appeared to be reduced. After 115 days of operation, surface ponding continued at 17 min. past the cessation of a high-level dose and 16 min. elapsed before 50 percent of the applied dose flowed out of the filter. The filter surface remained ponded for extended periods of time after 135 days of operation, and by day 149, continuous ponding persisted. At that point, loading to the filter was terminated.

A combination of surface raking and hydrogen peroxide treatment was used to rejuvenate the filter. Immediately after termination of the filter run, approximately 50 percent of the ponded surface was raked vigorously to a depth of 15 cm. The 7 to 10 cm of ponded water slowly drained through the filter during the next 30 min. The unraked sand surface was black and malodorous and covered with a thin (1 to 2 mm) mat of black solids. The raked area indicated a black discoloration to approximately 10 cm into the bed. After smoothing the sand bed, 5.7 liters of 15 percent H_2O_2 were dosed on the filter and flushed with water. The filter was placed back into service approximately 45 min. after this treatment.

Twenty-four hours after reloading the filter, the outflow rate was measured after a high-level dose had been applied. The surface remained ponded for almost 10 min. past cessation of the dose and 50 percent of the applied dose volume was collected as outflow after 30 min. Thus, the hydraulic conductivity of the sand bed had apparently not been significantly rejuvenated by the raking and the level of H_2O_2 treatment employed. This was supported by the fact that a succeeding filter run lasted only 45 days until continuous ponding reoccurred.

During this phase of operation, the influent wastewater temperature was relatively constant at 8 to 10 ^{o}C, while the air temperature in the filter head space slowly decreased from approximately 20 to 10 ^{o}C. To prevent freezing and maintain temperatures conducive for bacterial activity, supplementary heat was applied to the filter head space after about 100 days of operation.

The effluent quality achieved with the sand filter during phase II is summarized in Table 8. Effluent quality was consistently high and nearly complete nitrification was observed. As was noted in phase I, significant quantities of nitrogen were removed from the graywater by the filter. Cores taken within the top 20 cm of the bed indicated that approximately 12 percent of the applied nitrogen was retained in that filter volume. If the nitrogen distribution associated with the sand media were similar throughout the entire bed, all of the nitrogen removed could be accounted for. Further work needs to be done to establish the degree of denitrification that may occur in intermittent filters.

Table 8. Reduction in Mean Pollutant Concentrations in Graywater by Sand Filtration - Phase II*.

Parameter	Units	Septic tank effluent	Sand filter effluent**	% reduction
BOD_5	mg/liter	216	6	97
COD	mg/liter	528	112	78
TSS	mg/liter	60	10	83
TVSS	mg/liter	46	5	89
TN	mg-N/liter	15.5	8.9	43
Turbidity	NTU	54	2	96
Total coliforms†	log no./liter	7.84	6.34	88
Fecal coliforms†	log no./liter	6.91	4.78	97
Period	---	7-22-80 to 1-6-81	7-22-80 to 11-5-80	---

*Based upon analyses of 24-hr flow-composited samples.
**Effluent produced through intermittent filtration through 27 inches of coarse sand (E.S. = 1.37 mm, U.C. = 1.30) at a mean daily loading of 9.0 gpd/ft^2.
†Log-normalized data.

The removal of fecal indicators during the phase II study was approximately two logs, but concentrations in the filter effluent remained high. In fact, fecal indicator concentrations from intermittent filters were in the same range as those found for total residential wastewater following filtration (10).

Sand Filtration - Column Studies

The sand filter column studies were initiated in July, 1980. The length of filter runs for each column is presented in Table 9 along with average dosing frequency and hydraulic loading. Within one month of start-up, all medium sand columns were continuously ponded with influent graywater. The long filter run noted for column 4 was due to short circuiting. No reason for the premature clogging of column 5 could be delineated. The high loaded, coarse sand columns demonstrated some ponding after 220 days, but no ponding was observed over the 250 day study with the low loaded coarse sand units.

Table 9. Filter Run and Dosing Characteristics of Sand Filter Columns.

Column	Code[a]	First Ponding	First Overflow	Termination[b]	Dosing Frequency (dose/d)	Hydraulic Load (cm/d)	Quantity of Effluent (cm)
1	++	220	**	> 250	23.4	31.1	7783
2	++	220	**	> 250	23.4	31.1	7783
3	-+	8	34	38	21.0	28.0	1063
4	-+	8	51	86	22.3	29.7	2559
5	--	11	77	86	22.3	14.2	1223
6	--	15	37	43	21.6	13.8	592
7	+-	*	**	> 250	23.4	14.9	3720
8	+-	*	**	> 250	23.4	14.9	3720
Field Filter Phase II(++)		120	-	149	22.5	39.6	5680

(a) Media size: Loading (Table 2); (b) Terminated columns after 2000 ml overflow

* - no ponding up to 250 days

** - no overflow up to 250 days

During the first 80 days of the study, the column air temperature ranged from 20 to 28 °C promoting rapid maturation of the filters. When ambient temperatures dropped below 12 °C, supplementary heat was supplied to the columns. During the entire 250 day period, column air temperatures ranged from 4 to 28 °C. The graywater septic tank effluent temperatures varied between 7 and 12 °C in this same period.

The effluent quality obtained during the column studies is delineated in Table 10. The effects of sand size and hydraulic loading on effluent quality are not pronounced except in the case of the high loaded medium sand columns. Nitrification was prevalent in all but the high loaded medium sand, but was greatest in the low loaded coarse sand filters. From 26 to 55 percent of the nitrogen applied to the sand columns was not accounted for in the effluent. Similar nitrogen removals were described earlier in the field studies. Two to three log reductions in fecal

Table 10. Mean Effluent Concentration of Selected Pollutants by Sand Filtration-Column Study*.

UNIT	CODE[a]	BOD_5	TSS	VSS	TKN	MH_4-N	NO_2/NO_3	FECAL COLI
Septic Tank Effluent	-	243	58	46	15.2	6.6	0.3	6.90
1	++	18	15	7	3.5	0.8	3.9	4.38
2	++	13	13	6	3.7	1.0	3.4	4.48
3	-+	54	27	11	10.6	6.3	0.3	4.43
4	-+	60	39	24	12.0	6.9	0.4	5.08
5	--	15	24	13	8.3	4.1	3.0	4.89
6	--	14	10	3	7.0	2.2	3.0	3.00
7	+-	13	11	6	2.4	0.3	6.5	4.34
8	+-	10	11	7	2.0	0.3	5.8	4.20
Sand Filter (Phase II,++)		7	10	5	3.0	0.3	5.9	4.78

*All mean concentrations, mg/l, except fecal coliforms log no/liter; mean values collected over entire test period.

(a) Media size: Loading (Table 2)

coliforms through the filters still resulted in effluent concentrations similar to those found for sand filtered residential wastewater.

Maturation of the sand filter columns was essentially complete within 40 days of start-up. Prior to that time, effluent quality was variable and, generally, poor. Little nitrification was noted until after approximately 40 days.

DISCUSSIONS AND CONCLUSIONS

The results of this study, in conjunction with studies conducted by others, indicate that residential graywater contained substantial amounts of pollutants, including BOD_5, suspended solids, nitrogen, and phosphorus. Kitchen sink wastewater in the graywater stream was a major contributor to the high levels of these pollutants. When compared to total residential wastewater, graywater from all household basins, sinks, and appliances had lower concentrations of most conventional pollutants, with the exception of BOD_5. If kitchen sink wastes are excluded, graywater pollutant concentrations are significantly lower than in the total residential waste stream. The levels of fecal indicators in graywater were also found to be high, suggesting that it may be routinely contaminated with fecal material. If fecal material were the sole source of this contamination, approximately 1.2 g of wet feces would have to be contributed to the graywater stream in a 24 hour period. The true source of fecal coliforms in graywater and the disease potential of graywater is still unknown, however.

Graywater may be treated effectively by employing a septic tank-intermittent sand filter system. Effluent quality in a properly designed system is excellent, but fecal indicator concentrations remain high. Of significant interest is the length of filter run achievable with a coarse grained sand. Filter runs of less than 150 days were observed in the field tests at application rates of graywater of 39 cm/d dosed at one hour intervals. Sand column experiments with this same wastewater suggested that longer filter runs were achievable at lower loading rates on a similar coarse grained sand. The scale-up of these column results to full scale design is uncertain, however.

Based on substantial experience with the intermittent filtration of total residential wastes (8, 10, 13, 14), the results of this work raise some serious questions as to the value of residential waste segregation/graywater treatment if high quality effluents are sought. The graywater treatment system described herein does not appear to offer a substantial improvement in cost savings, effluent quality, or manageability over a similar system for total residential waste. Additional work is underway to further substantiate the findings of this study and to assess objectively the overall cost and management effectiveness of waste segregation practice under a variety of user constraints.

ACKNOWLEDGEMENTS

This study was supported in part by funds from the State of Wisconsin administered by the Small Scale Waste Management Project at the University of Wisconsin-Madison. Substantial funding was also provided by U.S. Department of Interior, Office of Water Research and Technology, Project Number OWRT B-10J-WIS.

REFERENCES

1. Olsson, E., L. Karlgren & V. Tullander: 1968, Household Wastewater. National Swedish Institute for Building Research, Stockholm, Sweden.

2. Ligman, K., N. Hutzler & W.C. Boyle: 1974, "Household Wastewater Characterization". J. Environ. Engr. Div., Amer. Soc. Civil Engr., 150, pp. 201-213.

3. Bennett, E.R. & E.K. Linstedt: 1975, Individual Home Wastewater Characterization and Treatment. Completion Report Series No. 66, Environmental Resources Center, Colorado State Univ., Boulder, Colo., 145 pp.

4. Laak, R.: 1975, "Relative Pollution Strengths of Undiluted Waste Materials Discharged in Households and the Dilution Waters Used for Each". Manual of Greywater Treatment Practice - Part II, Monogram Industries, Inc., Santa Monica, California.

5. Siegrist, R.L., M. Witt & W.C. Boyle: 1976, "Characteristics of Rural Household Wastewater". J. Env. Engr. Div., Amer. Soc. Civil Engr., 102, pp. 553-548.

6. Siegrist, R.L., W.C. Boyle & D.L. Anderson: 1981, A Field Evaluation of Selected Water Conservation and Wastewater Reduction System for Residential Applications. Technical Report WIS WRC81-02, Water Resources Center, University of Wisconsin, Madison, WI. 119 pp.

7. Saw, Chin Shuan: 1981, Characterization of Residential Greywater and Intermittent Filtration of Greywater Septic Tank Effluent through Sand Columns. M.S. Thesis, Dept. of Civil & Environmental Engineering, Univ. of Wisconsin, Madison, WI. 191 pp.

8. Hines, M. & R.E. Favreau: 1974, "Recirculating Sand Filters: An Alternative to Traditional Sewage Absorption Systems". Proceedings of National Home Sewage Disposal Symp. Amer. Soc. Agricultural Engr., Chicago, Ill, pp. 130-137.

9. Furman, J.T., W.T. Calaway & G.R. Grantham: 1955, "Intermittent Sand Filters-Multiple Loadings". Sew. and Ind. Wastes, 27, (3), pp. 261-276.

10. U.S. Environmental Protection Agency: 1980, Design Manual - On Site Wastewater Treatment and Disposal Systems. Office of Research & Development, MERL, EPA 625/1-80-012, 391 pp.

11. Brandes, M.: "Characteristics of Effluents from Separate Septic Tanks Treating Greyed Black Waters from the Same House". J. Wat. Poll. Cont. Fed., 50, pp. 2547-2559.

12. Kristiansen, R. & N. Skaarer: 1979, "BOV-vannets Sammensetning og Mengde. Undersøkelser i Danskerudfeltet på Ås". VANN, (2), pp. 1-6.

13. Small Scale Waste Management Project: 1978, Management of Small Waste Flows. USEPA, EPA 600/2-78-173.

14. Saver, D.K., W.C. Boyle & R.J. Otis: 1976, "Intermittent Sand Filtration of Household Wastewater". J. Environ. Engr. Div. Amer. Soc. Civil Engr., 102, pp. 789-803.

SESSION IX

STATE REGULATIONS AND POLICY REGARDING ON-SITE WASTEWATER DISPOSAL SYSTEM

Chairman: H. Ødegaard

REGULATIONS AND POLICY REGARDING ON-SITE WASTEWATER DISPOSAL SYSTEMS

James F. Kreissl

Sanitary Engineer/Program Manager
Urban Systems Management Section, Municipal Environmental Research Laboratory, U.S. Environmental Protection Agency, Cincinnati, Ohio 45268, USA

BACKGROUND

There are no Federal Environmental Protection Agency regulations regarding the design of on-site wastewater systems in the United States. In 1957 the U.S. Public Health Service published a document entitled "Manual of Septic Tank Practice" (1) which represented the first major compilation of post-World War II research. Prior to that time the states had administered a wide variety of codes, almost as many and varied as the number of states involved in their administration. The Manual of Septic Tank Practice (MSTP) had such an influence on the various codes that by 1971, a study of state septic tank-soil absorption system (ST-SAS) codes showed that modal values for ST-SAS design were identical to the MSTP (2). However, as the next decade unfolded, an unofficial survey of state codes revealed that all but three of the responding 47 states had revised their conventional ST-SAS regulations (3). This discussion will focus on the trends apparent in the changes in state regulations and their interpretation in light of recent research studies following a brief overview of how these regulations are administered.

A. S. Eikum and R. W. Seabloom (eds.), Alternative Wastewater Treatment, 303–320.

ADMINISTRATION

Ward (4) recently discussed the administration of on-site system technology regulations noting that eight states retain the sole permit issuing authority, while thirteen designate that role to the local unit of government. The remaining twenty-nine states share the administration of regulation. Comparing these figures to a 1975 study by Plews (5) indicates that the relative distribution in these categories has changed little in the past five years, i.e., the majority of regulation occurs at the local level, either exclusively or in concert with a state agency. Other studies verify the spectrum of control schemes among the 50 states, describing the various methods of code creation and enforcement (2, 6). Both Stewart (6) and Patterson et al. (2) favor strong state enforcement programs, while Plews (5) favors local enforcement of state programs. In the former case, the most persuasive argument is reduced political pressure at the state level, while in the latter it is clearer understanding of local conditions. Plews describes the difficulty in making the above determination by subdividing responsibilities with regard to the relative capabilities of both governmental levels.

The foregoing discussion of preferable governmental regulation is limited to conventional governmental units. In recent years several on-site system management approaches have been attempted using any one or a combination of the following (7):

1. Public agencies - state, county and city governments, as well as planning agencies and conservation districts.
2. Special service agencies - water, sewer and sanitary districts.
3. Private sector entities - private contractors, private utilities companies, rural cooperatives and property owner associations.

These management approaches are often quite successful in overcoming the significant historical shortcomings of reactive enforcement by existing regulatory agencies of the state and local government. The optimum management system is a function of the state enabling legislation, the on-site technology type and mix in the area, and the monitoring requirements imposed. The major management functions which must be satisfied are (7):

1. Planning
2. Site evaluation
3. System design
4. Installation supervision
5. Operation and maintenance
6. Financing
7. Water quality monitoring
8. Systems inspection
9. Public education
10. Program coordination.

In each community a unique optimum management solution should be developed which employs one or more of the public and private sector entities described above.

STATE CODE EVOLUTION

When Weibel et al. (8) surveyed existing state codes in 1947, the variation in requirements was quite immense. This was true for both septic tank and soil absorption system designs. Their survey accompanied a major research study which still has a significant influence on modern practice. The format of state ST-SAS codes is somewhat uniform in that it usually specifies such things as septic tank minimum sizes and internal appurtenances and soil absorption system location requirements, trench design specifications, and soil limitations. In describing the results of the Weibel et al. (8) survey and subsequent surveys (2, 3, 5), the results were often converted into simplified categories traditionally used in codes. These categories are often not fully descriptive of the code contents, but simplify comparative analysis. In the past decade state codes have become more complex, reflecting special site conditions and alternative systems to overcome those conditions which would otherwise preclude conventional ST-SAS installation. Therefore, trends and other qualitative information are reasonably extracted, but exact, quantitative information requires more thorough analysis of each code.

Several trends can be identified in reviewing the surveys of 1947, 1971, 1975 and 1980 (2, 3, 5, 8). In 1947, the minimum volume for a septic tank was specified as 1.9 m^3 (500 gal) by 21 of 47 states, while in 1980, 28 of 49 states required a minimum of 2.8 m^3 (750 gal), and 14 required 3.8 m^3 (1000 gal) or more. With relation to other tank specifications, the 1947 survey noted that 40 states specified a minimum length:width ratio, and 72 percent of those required that ratio be between 2 and 3. In 1971, only 8 states specified a minimum length:width, and all but one required it to be 2 to 3. Similarly, of the 41 states requiring a minimum depth, 89 percent required 1.2 m (4 ft) or more. In 1971 only 22 states had a minimum depth requirement, and over one-half of those required a minimum depth of 0.9 to 1.2 m (3 to 4 ft). Because of its more comprehensive nature, the 1947 survey contained information on many more details of tank requirements than the subsequent surveys. It noted, for example, that six states recommended two-compartment tanks, while the author is only aware of one state doing so at present. Also, in 1947 four states recommended against inlet baffling or submergence, a nearly universal requirement today.

What do the above trends seem to indicate? Clearly, the code-makers feel that larger tanks are advantageous. However, they are becoming less restrictive on the shape of the tank.

Given the reduced septic tank pumping frequency, requirement of the larger tank and the concerns for both increased wastewater flow due to modern fixtures and widespread use of garbage grinders, this progression would appear to be logical. However, earlier intent to improve tank performance through better internal design has apparently given way to perceived benefits of larger tanks.

It was noted in 1947 (8) that fourteen states utilized the percolation test for site evaluation and design. After the advent of the MSTP (1), the 1971 survey (2) revealed that twenty-six states required percolation testing for system design. By 1975, thirty-six of thirty-nine reporting states used the percolation test, but only 19 used the test results alone for system sizing (5). In 1980, thirty-two states specified its use in design (3). The percolation test itself is not standardized. Several investigators have demonstrated its extreme variability when run even in an approved manner. It has been proven that the test bears little relationship to a functioning soil absorption system in terms of actual soil hydraulics (9).

In analyzing the above trends, the percolation test appears to remain extremely popular, despite its proven shortcomings. Since the test appears to be straightforward, semi-skilled technicians appear able to perform it with some reasonable level of proficiency. Despite the fact that tests have been developed which actually measure hydraulic conductivity of the soil under a variety of moisture conditions, the percolation test remains in general use because of the complexity of these tests and a lack of information on how to use the more accurate data. Therefore, the attitude of several more progressive code-writers has been to either continue using the percolation test only as one of several data inputs to system design, or to eliminate it altogether and rely on other site evaluation data. The other data in both of the above cases include detailed soil analysis of texture, depth and a structure, landscape location and drainage, and prior experience with the identified soil for on-site wastewater disposal. Unfortunately, relatively few states fall into this category, but the "blind" use of percolation tests for sizing of soil absorption systems does appear to be waning.

Some interesting design trends for absorption trenches are also apparent in these reviews. Trench dimensions have been subject to debate for decades. In wetter eastern states, it has been noted that shallower, wider trenches are sized on the basis of bottom infiltration area, while drier western states employ narrower deeper trenches sized on the basis of sidewall infiltration area (12). The results of the 1980 survey show the difference in average dimensional requirements to be negligible. The 1947 survey revealed that the modal minimum width requirement was 45 cm (18 in), while in 1980 it was 30 cm (12 in). With regard to minimum depth, modal values for 1947 and 1980 were identical at 60 cm (24 in), with few revisions over the 33-year period. Interpretation of these trench dimensional requirements is difficult, but they could reflect a belief in the use of sidewall absorption in favor of bottom area or merely reflect changes in equipment employed for digging trenches.

Similarly, the required distance between trenches in 1947 was 2.1 m (7 ft), while in 1980 it was 1.8 m (6 ft). This reduction may be due to concern over larger overall SAS size requirements and setbacks, which will be dealt with later. By reducing trench separation requirements, more absorption area can be provided within a given land area. Since the physical reason for spacing trenches has not been established other than that dictated by construction equipment dimensions, this modest reduction does not appear to create interfering flow patterns between trenches, but would increase overall areal loading by almost 15 percent.

One very drastic change in trench requirements is in the minimum cover required over the trench. In 1947 half of the codes surveyed cited a "preferred" minimum cover of 45 cm (18 in), while in 1980 more than half of the states required only 15 cm (6 in). Clearly, the oft-stated concern for freezing and mechanical damage has diminished, and possibly, some attempt to more effectively use evapotranspiration assistance in disposal may be apparent.

Design loadings determined in the 1947 survey were expressed in a variety of ways in different states. For example, some codes expressed required soil absorption in trench bottom area per capita, while others employed length of trench per capita and bottom area per bedroom. Without other specific details of the various codes direct comparisons are impossible. A few items of note included the general acceptance of clays as being unsatisfactory for soil disposal and the absorption area requirement range of from less than 9.3 m^2 (100 sq ft) for sandy soils to 93 m^2 (1000 sq ft) for the finest-textured acceptable soils. In those states which had already encoded the percolation test in 1947 absorption area requirements had doubled by 1971 for sandy soils along with significant, but smaller, percentage increases for finer-grained soils with marginally acceptable percolation times. With regard to very permeable soils, the 1975 survey showed nine states with minimum SAS sizes, no matter what results the percolation test might provide. The 1980 survey indicated that this number had grown to 29 states, but that the minimum sizes varied widely, from 6.5 m^2 (70 sq ft) to 46.5 m^2 (500 sq ft), with a mean minimum size of 16.3 m^2 (175 sq ft).

The obvious conclusion which can be made from these changes, is that the required sizing of conventional soil absorption fields has generally increased over the intervening period. The reason for these increases are probably due to unsatisfactory performance of earlier, smaller systems. Given the overall lack of understanding of the design and functioning of the SAS and its uncertain relationship to the perc test, the logical choice was to increase the required absorption area. Given that numerous studies have shown that equilibrium rates of infiltration do occur with time, it is feasible that larger and larger systems could eventually permit longer system service life. Such reasoning, however, neglects two important factors. The first is economics, which includes not only the extra cost of construction of larger beds, but also, when combined with required setback or separation distances, the cost of larger parcels of property to permit construction of these systems. The second factor is purification of wastewater to protect groundwater from contamination. Without control over the method of wastewater application, the potential for groundwater contamination is high during the early stages of operation, and this period of high contamination potential is merely lengthened by larger systems without some form of improved distribution.

The two aspects of codes which deal with groundwater contamination are the vertical separation distance to groundwater and the horizontal separation or setback requirements to household potable water supplies. The vertical separation requirement provides unsaturated soil "travel time" to accomplish treatment of the wastewater before it enters the groundwater. The horizontal separation is commonly perceived to provide further treatment time in the saturated zone before any human consumption of the treated wastewater/groundwater mixture can occur.

The 1947 survey showed 34 states with vertical separation requirements varying from 0 to 1.2 m (0 to 4 ft), with a mode of 1.2 m (4 ft). In 1980, 41 states required the same range and mode, however, the average separation requirement had risen from 0.8 m (2.6 ft) to 0.9 m (3.0 ft). Of 20 directly comparable states, 14 had not changed their requirements, 4 increased them, and two had decreased them. Since the contact time between the wastewater and unsaturated soil is the most effective barrier against groundwater contamination, design requirements must insure that no short-circuiting can occur which would allow insufficient treatment. Bouma (9) has described the contact time concept in detail. By integration of the Darcy equation he demonstrated the need to provide a minimum of 0.9 m (3 ft) of unsaturated soil between the trench bottom and the top of the saturated layer to maintain hydraulic acceptance rates at design equilibrium conditions. Contact times for sandy (unstructured) soils can be determined readily through knowledge of nominal hydraulic loadings and moisture tension conditions because of piston flow in these soils. Structured soils are far more complex and often short-circuit wastewater through natural macropores or those created by roots and worms. Bouma (9) presents examples of this short-circuiting to show nominal contact time to actual contact time ratios in some soils of 23:1 or more. Therefore, the simple designation of a vertical separation distance may not ensure against the passage of pathogens to groundwater in certain soils, especially if no control of hydraulic loading is provided. However, a requirement of 0.9 m (3 ft) or more of vertical separation would appear to be a reasonable attempt to improve purification. Unfortunately, 13 of 41 states required less than 0.9 m (3 ft) of separation in the 1980 survey.

Horizontal separation or setback distances to water supplies, which in most rural locations are wells, also varied widely in the surveys. In 1947 the mode (17 of 32 states) setback distance required between the SAS and the water supply was 15 m (50 ft), while in 1980 it was 30 m (100 ft) in 21 of 38 states. In relation to setback requirements for surface waters both surveys indicate that modal separation was 15 m (50 ft). The water supply setback requirement increase would appear to reflect a concern for the travel of pathogens through the groundwater to a well supply.

In reviewing the waterborne disease outbreak data for 1970 to 1974, the author (10) noted that in all outbreaks traced to ST-SAS's, the separation distances were in excess of 30 m (100 ft). Similarly, Ford et al. (11) statistically analyzed well contamination by nitrates in Colorado to show the probability of nitrate-nitrogen contamination (>10 mg/l) to be 21.8 percent at 30 m (100 ft) and 9.4 percent at 61 m (200 ft). In all of these cases, the data are influenced by several factors in addition to separation distance alone. However, the mere codification of a minimum setback distance in and of itself does not insure against pathogenic contamination of wells, nor does it prevent the passage of nitrates to wells and surface waters.

Several other miscellaneous factors have crept into state codes in recent years. Among them are replacement area requirements, reduced infiltrative areas for alternative pretreatment systems, and provisions for alternative systems. The 1980 survey indicated that 21 of 45 states require an SAS replacement area (generally, 100 percent of the original system area). Such requirements, without any alternating bed provisions, appear to admit both failure to present design or control criteria and a lack of understanding of system functioning.

In only four states included in the 1980 survey are reductions in infiltration area permitted where aerobic treatment units are employed in place of septic tanks. The maximum reduction is 33 percent in the usual range of acceptable percolation rates. In one state the allowable percolation rate is extended from 24 to 48 min/cm (60 to 120 min/in) for aerobic pretreatment. No scientific evidence supports any reductions in the finer soils, although some evidence does exist to support size reduction in sandy soils.

Provision for alternative designs has been incorporated into 48 of the 50 state codes as of the 1980 survey. However, the author is aware of alternative systems in both of the states whose codes apparently do not allow them. Therefore, the need to provide alternative systems which can overcome site restrictions which preclude conventional ST-SAS's is apparently universal.

CODE ELEMENTS AND PURPOSES

Plews (5) has presented a historical picture of state and local on-site system code development, describing their disease prevention origins and their evolution into land use controls, growth stimulation or limitation mechanisms, and other political goals.

To properly accomplish these goals a thorough evaluation of the proposed site is required to determine optimum landscape location, soil texture and structure analysis with depth, groundwater and bedrock data, hydraulic conductivity data, and information on the volume and character of the wastewater to be absorbed by the soil. The ability to obtain the above data is limited in most areas of the U.S. However, some states and localities have made great advances in the form of upgrading site evaluation through requirements that more highly qualified personnel perform these analyses. In these instances the associated codes are less stringent, giving qualified people an opportunity to consider a variety of designs which may be optimal to each site. In the majority of cases, however, less qualified personnel are employed to determine site suitability. Therefore, codes in these states are generally quite rigid, permitting minimal adjustments of designs to site-specific conditions. In many of these same jurisdictions, when a site is declared unsuitable by those personnel, professional engineers are permitted to propose designs which can overcome site limitations. Unfortunately, few engineers are properly educated to ably provide this type of advanced design.

Since all codes are based on the premise of protecting public health and no present technique of on-site pretreatment for soil disposal is capable of complete pathogen removal, the health protection factor centers on the proper siting, design, and construction of the soil absorption system (10). The two primary routes of human disease transmission are: 1) passage of pathogens or chemical contaminants from the disposal trench downward to the groundwater, travelling therein to a point of extraction in sufficient quantity to cause disease, and 2) surfacing of unpurified wastewater where direct human contact or overland runoff to improperly cased wells can occur. In order to avoid the latter, the system must be properly sited, designed, and constructed to accept the entire volume of wastewater for a long period of time. To avoid the former, the system must be properly designed to avoid localized overloading and to provide sufficient contact time in the unsaturated soil zone to ensure purification.

Code use of horizontal separation distance or setback requirements is generally considered their primary protection against passage of pathogens to nearby water supplies. Although the primary protection against this occurrence lies in the purification afforded in the unsaturated zone which is determined by vertical separation distances from trench bottoms to groundwater tables, this fact is not often recognized by those responsible for code development. Depending on the type of aquifer, the distance of pathogen travel could easily exceed any reasonable horizontal setback distance and be limited only by their ability to survive under the physical conditions of the groundwater. Therefore, the increased setback distances noted earlier in newer codes are relatively ineffective without ensuring sufficient contact time in the unsaturated soil prior to introduction to the groundwater.

Previously, the vertical separation requirements of newer codes were noted to be slightly increased over the older ones.

However, in and of itself this distance may not provide the necessary contact time. In those soils where minimum separations are marginally available and conventional gravity loading of the trench is employed, there is a high probability of groundwater contamination in coarser soils and fine, structured soils with macropores due to natural conditions, root channels, or worm activity. This probability is highest during the early months of operation prior to the development of a clogging mat or crust. The problem results from poor distribution of the relatively small wastewater generating events common to individual homes. With poor distribution, as illustrated in Figure 1, the localized overloading may be several times the design loading, resulting in extensive vertical penetration through large soil pores of pathogen-laden wastewater. Once the crust developes, it shuts off the larger pores, limiting conduct of liquid to the smaller pores where longer detention times are assured. In order to avoid these conditions in the soils described above, improved distribution must be encoded in addition to proper vertical separation.

To avoid surfacing of unpurified wastewater, the other major health concern, the system must be sited, designed, and constructed to accept the wastewater generated. The majority of the other code requirements are supposed to deal with proper design of absorption systems in accordance with the code-creators' concepts of proper design and, to some degree, knowledge of equipment and procedures used in construction. Because of a significant difference in design recommendations between the most influential researchers of the past few decades, a variety of codes exist. Otis et al. (12) have compared the various trench design parameters, with respect to the major schools of thought. Because each has performed their studies under the conditions and soils of their region, a certain degree of difference may be inherent in those differences, particularly where empirical relationships are concerned.

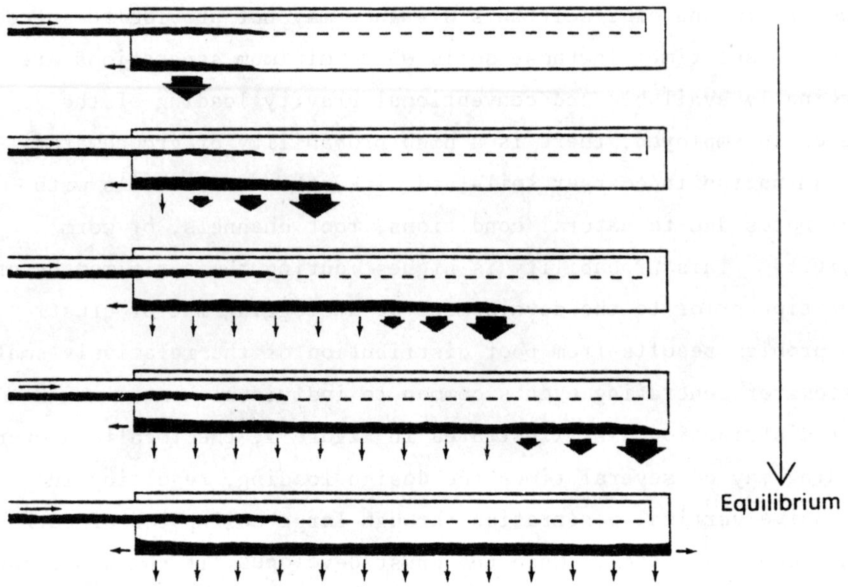

Figure 1. Hydraulic regime in a conventional gravity loaded trench system during the initial months on operation.

NATIONAL AND REGIONAL GUIDANCE

Some national and regional documents which provide guidance for on-site system design have also influenced the state-of-the-art, though not as directly as the state and local codes. The U.S. Departments of Housing and Urban Development and Agriculture (Farmers Home Administration) have produced a set of standards, revised in 1973 (13). In essence these standards are designed for use where local standards are non-existent. Septic tank requirements and percolation test-system sizing requirements are about the same as the MSTP (1). Other absorption system criteria are generally similar to the MSTP, although the desire for more shallow trenches, less horizontal separation, and greater distance between parallel trenches is reflected. The Great Lakes-Upper Mississippi River Board of State Sanitary Engineers have recently (1980) issued a new set of recommended standards for on-site systems (14). More popularly known as "Ten States Standards", this document attempts to combine conventional and alternative on-site technologies in one set of standards. Noteworthy differences from the MSTP are reduced setback requirements, larger required septic tank volumes, wider trenches, and reduced vertical separation requirements to groundwater. Curious requirements include a limitation of 12 percent on the slope of the site, whereas state codes set limits from 15 to 50 percent. Several national plumbing codes also provide on-site system code information.

In 1980 the U.S. Environmental Protection Agency produced the Design Manual for On-Site Wastewater Treatment and Disposal Systems (15). This document is less prescriptive than those described above, which are very similar to most state codes. It is intended to assist site evaluators, designers, and regulators by describing systems which have performed successfully, under what circumstances, and the operation and maintenance experience for each. Although it addresses the same issues as the prescriptive codes, it does so with explanation of why various factors are codified, rather than by any required numerical value. In addition, the site evaluation procedures described are extremely valuable tools which are not normally provided in codes. Since this document deals with a variety of alternative systems, screening methods are provided to permit the designer to evaluate in detail only those systems (of those described in the text) that are appropriate to overcome the various site limitations, as shown in Table 1.

SUMMARY

No Federal Environmental Protection Agency standard code exists in the United States for control of on-site wastewater systems. However, state and local codes appear to be dynamic

Table 1. Selection of Disposal Methods under Various Site Constraints.

Method	Site Constraints											
	Soil Permeability			Depth to Bedrock			Depth to Water table		Slope			Small Lot Size
	Very Rapid	Rapid-Moderate	Slow-Very Slow	Shallow and Porous	Shallow and Nonporous	Deep	Shallow	Deep	0-5%	5-15%	15%	
Trenches		X	X[2]			X		X	X	X	X	X[4]
Beds		X				X		X	X			X
Pits		X				X		X	X	X	X	X
Mounds	X	X	X	X	X	X	X	X	X	X		
Fill Systems	X	X[1]	X[1]	X	X	X	X	X	X	X	X	X[4]
Sand-Lined Trenches or Beds	X	X	X[2]			X		X	X	X[3]	X[3]	X[4]
Artificially Drained Systems		X				X	X		X	X	X[3]	
Evaporation Infiltration Lagoons		X	X[5]			X		X	X			
Evaporation Lagoons (lined)[4,5]	X	X	X	X	X	X	X	X	X			
ET Beds or Trenches (lined)[4,5]	X	X	X	X	X	X	X	X	X	X[6]		
ETA Beds or Trenches[4]		X	X			X		X	X	X	X	X

[1] Only where surface soil can be stripped to expose sand or sandy loam material.

[2] Construct only during dry soil conditions. Use trench configuration only.

[3] Trenches only.

[4] Flow reduction suggested.

[5] High Evaporation potential required.

[6] Recommended for south-facing slopes only.

X means system can function effectively with that constraint.

with time, as shown by four surveys spanning 33 years. As noted by Plews (5) these codes do not always reflect only the status of technology, but often political decisions based on development pressures. Few states have incorporated updated information on the potential passage of pathogens to groundwater, but a lack of a significant number of disease outbreaks clearly traceable to on-site systems has provided little impetus for change. National guidelines do exist, but their impact has been minimal at this time. Finally, the value of any code is established primarily through the quality of staff enforcing that code, i.e., their understanding of the technical and management requirements for successful on-site system performance and their ability to work with local entities and individuals to accomplish local goals without jeopardizing public health and the environment.

REFERENCES

1. Manual of Septic Tank Practice, 1967. USPHS Publication No. 526.
2. Patterson, J.W., R.A. Minear & T.K. Nedved: 1971, "Septic Tanks and the Environment". NTIS Report No. IIEQ 71-2.
3. Virginia Department of Health, unpublished survey of state regulations, 1980.
4. Ward, R.D.: "Alternative Management Strategies for Colorado's Small Wastewater Flows", presented at 4th Workshop on Home Sewage Disposal, Fort Collins, Colorado.
5. Plews, G.D.: 1977, "The Adequacy and Uniformity of Regulations for On-Site Wastewater Disposal - A State Viewpoint", in Proceedings of National Conference on Less Costly Wastewater Treatment Systems for Small Communities, USEPA.
6. Stewart, D.E.: 1975, "Legal, Planning and Economic Considerations of On-Site Sewerage Systems". Proceedings of the National Home Sewage Disposal Symposium, ASAE Pub. PROC-175.
7. Wiswall, K.C. & P.A. Ciotoli: 1980, "Management of Alternative Systems: Issues, Problems, Constraints and Opportunities". Proceedings of 6th National Conference on Individual Onsite Wastewater Systems.

8. Weibel, S.R., C.P. Straub & J.R. Thoman: 1949, "Studies on Household Sewage Disposal Systems (Part I)". Federal Security Agency - Public Health Service.

9. Bouma, J.: 1977, "Soil Survey and the Study of Water in Unsaturated Soil". Soil Survey Paper No. 13, The Netherlands Soil Survey Institute.

10. Kreissl, J.F.: "Current Practices - Subsurface Disposal". Proceedings of Conference on Microbial Health Considerations of Soil Disposal of Domestic Wastewaters, USEPA Report, in press.

11. Ford, K.L., J.H. Schott & T.J. Keefe: 1980, "Mountain Residential Development Minimum Well Protective Distances Well Water Quality". Journal of Environmental Health, 43, No. 3, p. 130.

12. Otis, R.J., G.D. Plews & D.H. Patterson: 1978, "Design of Conventional Soil Absorption Trenches and Beds". In Proceedings of the Second National Home Sewage Treatment Symposium, ASAE Publication 5-77.

13. U.S. Department of Housing and Urban Development: 1978, "Minimum Property Standards for One and Two Family Dwellings, No. 4900. 1". Reprinted in: On-Site Wastewater Disposal, WVU Cooperative Extension Service, Pub. No. AP-170.

14. Great Lakes - Upper Mississippi River Board of State Sanitary Engineers: 1980, "Recommended Standards for Individual Sewage Systems".

15. U.S.EPA: 1980, "Design Manual Onsite Wastewater Treatment and Disposal Systems.

PRESENT NORWEGIAN PRACTICE AND ACTIVITIES FOR IMPROVEMENT OF ON-SITE WASTEWATER DISPOSAL

Lasse Bræin

Civil Engineer, Norwegian State Pollution
Control Authority (SFT), Oslo Norway

INTRODUCTION

This paper gives a brief survey of the Norwegian regulations and guidelines concerning on-site wastewater disposal. It presents some important experiences with the Norwegian practice and some aims and present activities to improve wastewater treatment and disposal in areas where municipal sewerage systems are not available.

REGULATIONS AND RECOMMENDATIONS

According to the Norwegian water pollution act (1970), it is necessary to have permission to pollute the physical environment. For municipal wastewater such permission may be given either by the county authorities or the local building authorities.

County authorities handle the applications for wastewater flows from more than 7 households. They will make certain demands concerning

- wastewater treatment
- effluent quality
- pipeline network
- maintenance.

A. S. Eikum and R. W. Seabloom (eds.), Alternative Wastewater Treatment, 321–331.

The Norwegian State Pollution Control Authority (SFT) has provided guidelines and recommendations for conventional sewerage systems and treatment plants, but as yet there are no guidelines for soil absorption systems, other than specifications for sizing, designing and constructing large septic tanks.

Local building authorities handle the applications for single households and cabins and groups of less than 7 such buildings.

The Ministry of Environment has established rules and guidelines for the local handling of such situations. The rules are as follows:

- For the settling of the applications
- the local responsibility for
    - the election of treatment system
    - the approval of a responsible person for the building
    - the control of building and maintenance.

The guidelines are as follows:

- Site evaluations
- location of system
- site criteria for dimensioning and design
- alternative systems
- control, check-lists.

These guidelines describe five alternative on-site solutions all based upon soil as the renovating medium.

If other systems are to be used, such as small package plants, permission must be obtained from the county authorities. However, for small flows (less than 50 persons) conventional treatment plants are quite rare. The alternative systems are shown in Figures 1-5.

The technical guidelines also recommend alternatives to water carriage systems when the conditions are not appropriate for an infiltration or sandfilter solution. Most common solutions are:

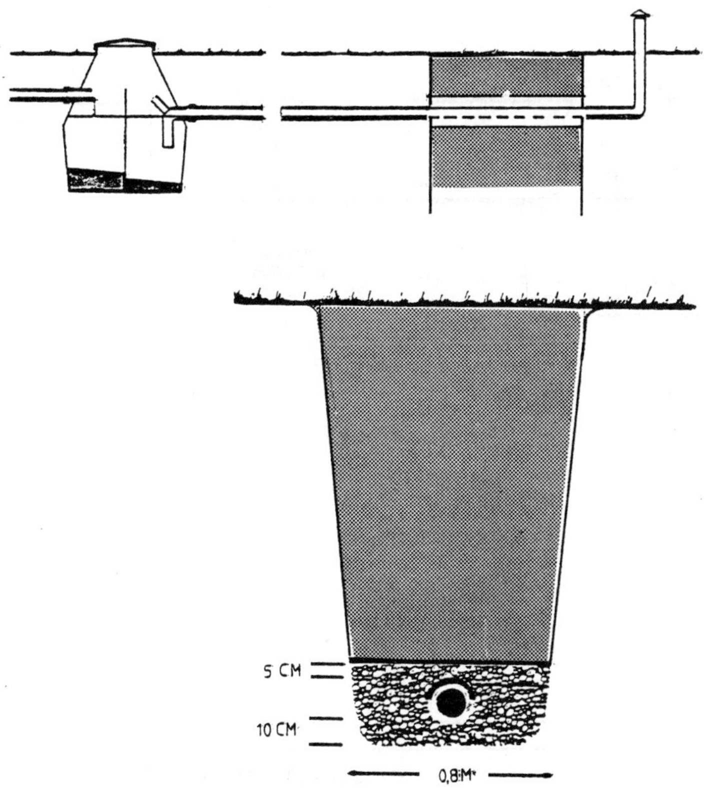

Figure 1. Infiltration Trench.

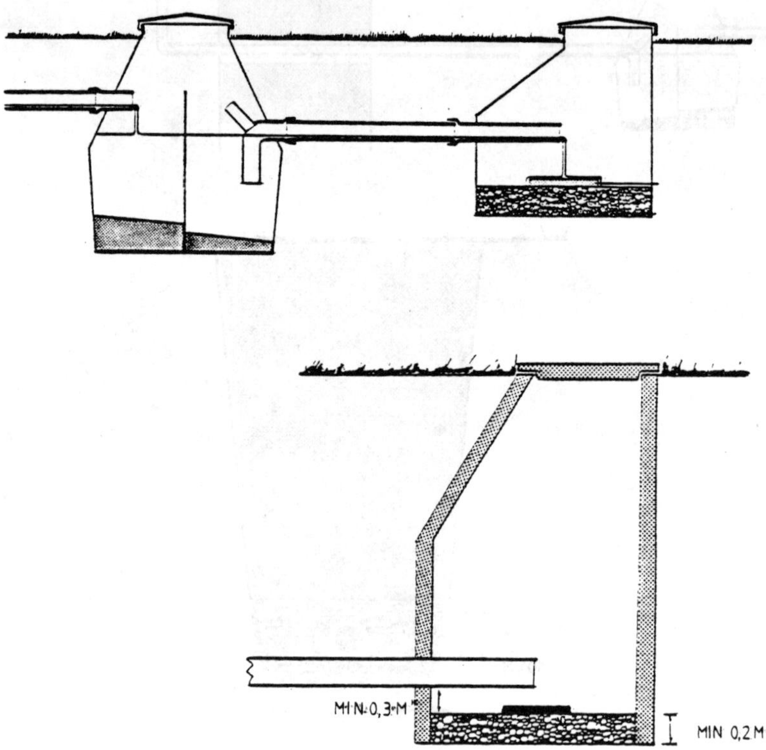

Figure 2. Seepage Pit (for Buildings without Water Closet only).

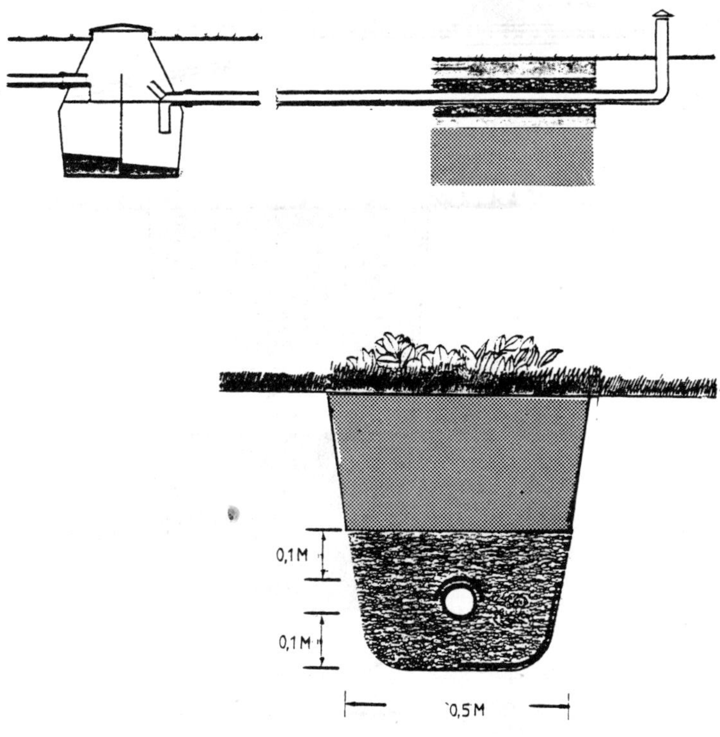

Figure 3. Evapotranspiration Trench
(for Cabins without Water Closet only).

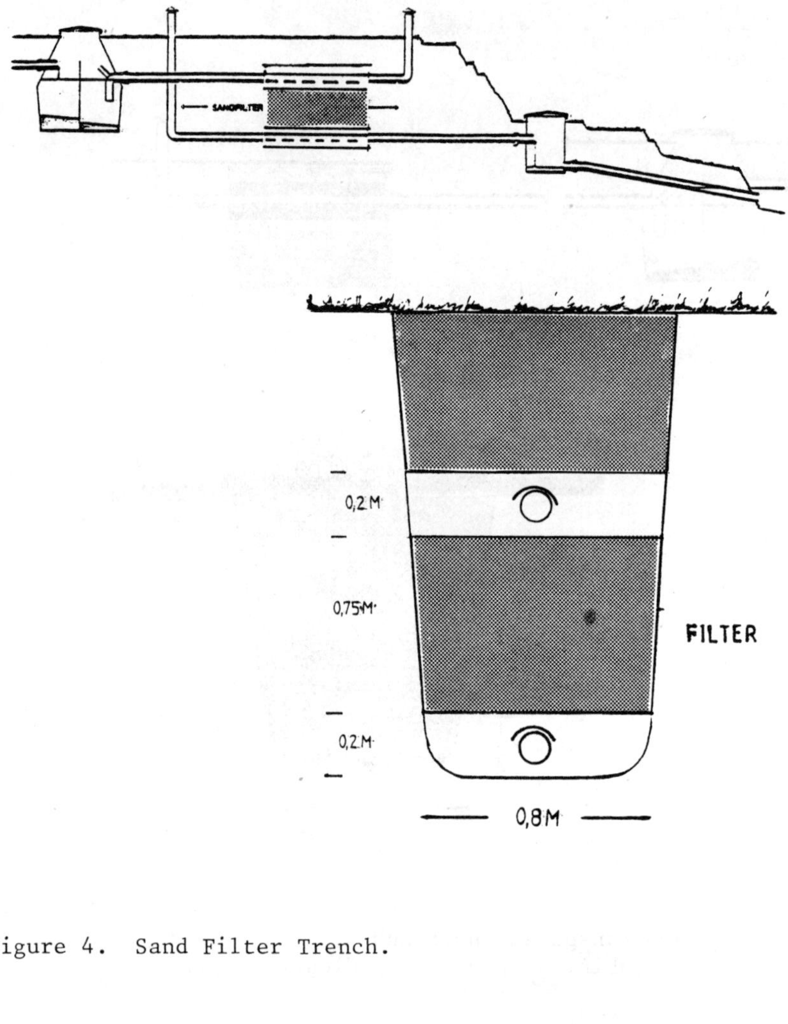

Figure 4. Sand Filter Trench.

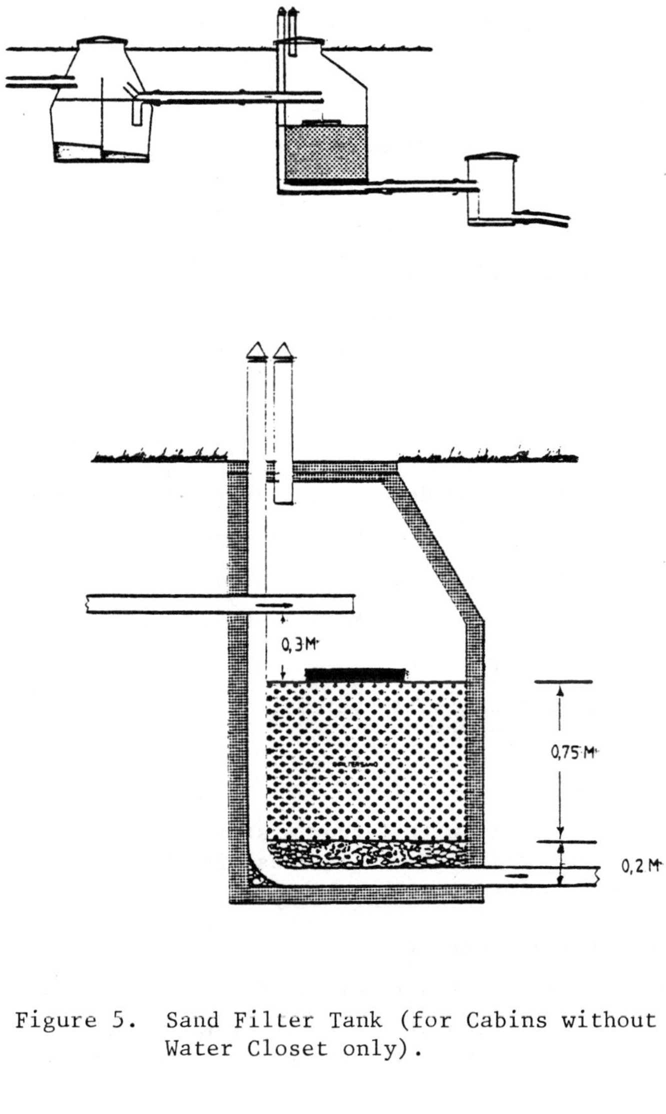

Figure 5. Sand Filter Tank (for Cabins without Water Closet only).

- Watersaving closets with a holding tank
- composting toilets.

The SFT has also set quality standards for composting toilets (presented by O. Molland in these proceedings). For septic tanks, quality standards are given for the materials, construction, and production in plastics, like

- polyester with glass fiber
- polyethylene.

The test according to these standards are done by "Det norske Veritas". Prefabricated concrete septic tanks are tested by Controlcouncil for Products of Concrete, according to national standards for prefabricated concrete products.

EXPERIENCES

It is believed that the described systems should perform satisfactorily if all the necessary precautions are taken. Inspections have shown, however, that in practice there are many problems with on-site solutions. Most systems have defects which are due to a variety of reasons:

I. Lack of knowledge about soil systems,
   - general uncertainty
   - misunderstandings
   - special uncertainty with larger systems and the capability of soil systems in areas with high density development.

II. Management problems consisting of
   - lack of capacity by the local authority
   - lack of qualification of the contractor or the official concerned with the project.

   This may lead to
   - very restrictive practice
   - pollution of drinking water supplies
   - destruction of systems

because of

- poor site evaluation
- selection of inadequate system
- improper construction
- lack of maintenance.

III. Technical problems like

- leakage from septic tanks
- insufficient loading equipment
- inadequate pipelines.

IV. There is a dearth of information for problem soils, areas with no soil at all, and areas with heavy restrictions.

V. There is a need for alternative solutions to the water closet for households and cabins. These solutions should be sufficiently attractive to represent a real alternative with respect to the user, the environment and to the municipality.

APPROACH TO AN IMPROVED NORWEGIAN PRACTICE

Since 1979 the SFT has been engaged in many activities to improve the Norwegian practice with on-site wastewater disposal systems. Through these activities attempts will be made to solve some of the problems previously described. A general aim for this work is that

> "there should be alternative solutions for wastewater treatment in rural areas in order that the pollution problem will not prevent an areal exploitation."

The new system should possess the following characteristics:

- stable reduction of pollutants as predicted
- long lifetime
- simplicity
- low capital costs
- low maintenance costs.

There should be appropriate alternatives to the water closet. SFT has granted about 300,000-400,000 NOK each year since 1979 to support research programs and to other projects.

The research program of Agricultural Research Council of Norway (NLVF) (presented by Rolv Kristiansen in these proceedings) and the research program of The National Swedish Environment Protection Board (SNV) (presented by Ulf von Brömssen in these proceedings) represent some important contributions to the understanding and knowledge about soil systems. In addition to the scientific reports generated from these programs, there will be users' manual convenient for people working on these problems.

One such manual will deal with the following topics:

- Site evaluation
- reduction of phosphorus
- hygienic aspects
- reduction of nitrogen
- clogging and hydraulic capacity
- frost
- effect upon ground water.

To implement the results from these research programs, a coordinating group of Swedish and Norwegian administrators and scientists has been formed. This group will produce design manuals for soil systems, and conduct studies of technical equipment like the septic tank and the pipelines to achieve the necessary flow distribution. There will be detailed manuals for small flow systems, less than 25 persons, and guidelines for larger soil systems, as well as functional standards for other small flow treatment plants.

Presently, studies are going on at the Water and Harbour Laboratory and the Technical University, both located in Trondheim, Norway.

To promote the development of appropriate technical equipment for on-site wastewater disposal in Norway, a group with members from interested industries, research institutions, and SFT has been formed. This organization provides a forum for exchange of information to the advantage of all parties, and even allows private enterprise to participate in the preparation of quality standards and tests.

The proper training of people involved in the practice of on-site wastewater disposal is another important requirement. To this end a four week course on pipeline construction, which includes on-site systems, has been developed. In addition, a training course for local authorities, which will concentrate on site evaluation, soil system functions, controls, etc. is under preparation.

Regular septage collection is important for a proper maintenance of on-site systems. The Norwegian guidelines recommend 4 m^3 septic tanks for each household with yearly septage collection. These recommendations are presently being reviewed, and it is anticipated that guidelines for municipalities concerning collection, transportation, treatment, and disposal of septage will be forthcoming.

Another problem that needs attention, is the clarification of when it is proper to use on-site wastewater disposal methods in "semi-rural" areas. Several areas known to the author have pollution problems, but the traditional conventional solutions are cost prohibitive, and yet it is not clear that on-site methods would be appropriate.

To solve these problems a cooperation with land use planners is needed. The writer hopes that through these efforts some of the identified problems will be solved, and the practice concerning on-site wastewater treatment systems will be improved.

LIST OF PARTICIPANTS

Mr. M. Andersson	The National Swedish Environment Protection Board, Institute of Microbiology S-750 07 Uppsala, Sweden
Mr. A. Berg	Norwegian Hydrodynamic Laboratories, Div. River and Harbour Laboratory P.O.Box 4118 Valentinlyst 7001 Trondheim, Norway
Dr. T. Bilstad	Rogaland Regional College P.O.Box 2540, 4001 Stavanger, Norway
Dr. W.C. Boyle	University of Wisconsin, Dept. of Civil & Environmental Eng. Madison, Wisconsin 53 706, USA
Mr. L. Bræin	Norwegian State Pollution Control Authority P.O.Box 8100-dep., Oslo 1, Norway
Dr. U. von Brömssen	The National Swedish Environment Protection Board P.O.Box 1302, 171 25 Solna, Sweden
Mr. E. Danielsen	Department of the Environment P.O.Box 8013-dep., Oslo 1, Norway
Dr. A.S. Eikum	Norwegian Institute for Water Research P.O.Box 333, Blindern, Oslo 3, Norway
Dr. C.G. Enfield	US Environmental Protection Agency Ada, Oklahoma 74820, USA
Mr. P. Engløv	VIAK AB, P.O.Box 452, 551 16 Jönköping, Sweden
Mr. S. Ensby	The Agricultural University of Norway 1432 Ås-NLH, Norway
Mr. T. Evjen	Biovac Norway A/S Bogstadveien 6, Oslo 3, Norway
Mr. O.F. Frederiksen	R. Brusletto A/S Lilleakerveien 31, Oslo 2, Norway

Mr. F.S. Furuset	Northern Øyeren Sewage Board (ANØ) P.O.Box 38, 2007 Kjeller, Norway
Mr. O.M. Grini	Norwegian State Pollution Control Authority P.O.Box 8100-dep. Oslo 1, Norway
Mr. K. Halvorsen	Bardu Municipality P.O.Box 53, 9250 Bardu, Norway
Mr. Chr. Harr	Trygve Elind A/S Sandviksveien 36, 1322 Høvik, Norway
Mr. S. Hoffner	National Bacteriological Laboratory S-105 01 Stockholm, Sweden
Mr. P.D. Jenssen	Agricultural Research Council of Norway The Agricultural University of Norway P.O.Box 9, 1432 Ås-NLH, Norway
Mr. R. Klubnes	Bergen College of Engineering Lars Hillesgt. 34, 5000 Bergen, Norway
Dr. T.-E. Korkman	Ministry of the Environment National Agency of Environmental Protection Strandgade 29, 1401 Copenhagen K Denmark
Mr. J.F. Kreissl	US Environmental Protection Agency 26 West St. Clair Street Cincinnati, Ohio 45268, USA
Dr. R. Kristiansen	Agricultural Research Council of Norway The Agricultural University of Norway P.O.Box 9, 1432 Ås-NLH, Norway
Prof. R. Laak	University of Connecticut Dept. of Civil Engineering U-37 Storrs, Connecticut 06268, USA
Mr. L.B. Langaard	The Water Quality Institute, ATV Agern Alle 11, 2970 Hørsholm, Denmark
Mr. A. Lind	The National Swedish Environment Protection Board P.O.Box 1302, S-171 25, Solna 1, Sweden
Mr. P. Lindbak	Fredriksborgveien 31 B, Oslo 2, Norway
Mr. K. Lohne	STATOIL P.O.Box 300, 4001 Stavanger, Norway
Mr. O. Molland	Agricultural Research Council of Norway The Agricultural University of Norway P.O.Box 9, 1432 Ås-NLH, Norway

Mrs. K. Mäkinen	Avaruuskatu 3 E 87 02210 Espoo 21, Finland
Mr. P. Nilsson	Technical College, Lund University P.O.Box 725, 200 07 Lund, Sweden
Mr. F.K.G. Nyberg	Water Resources Engineering Royal Institute of Technology Fack, S-100 44, Stockholm, Sweden
Mr. J. Ofte	Vidar Tveiten, Consulting Engineers 3840 Seljord, Norway
Mr. W. Olsen	Office of Vestfold Municipality Fylkeshuset, 3100 Tønsberg, Norway
Prof. R.J. Otis	University of Wisconsin Dept. of Civil & Environmental Eng. Madison, Wisconsin 53706, USA
Mr. B. Paulsrud	Norwegian Institute for Water Research P.O.Box 333, Blindern, Oslo 3, Norway
Mr. M. Posejpal	Vera Factories P.O.Box 233, 3201 Sandefjord, Norway
Mr. E. Santala	National Board of Waters P.O.Box 250, SF-00101 Helsinki 10 Finland
Prof. R.W. Seabloom	University of Washington Dept. of Civil Engineering FX-10 Environmental Engineering & Science Seattle, Washington 98195, USA
Mr. N. Skaarer	The Agricultural University of Norway Dept. of Landscape Architecture P.O.Box 23, 1432 Ås-NLH, Norway
Mr. M. Solgaard	Østlandskonsult A/S P.O.Box 141, 1601 Fredrikstad, Norway
Mr. T.A. Stenström	National Bacteriological Laboratory S-105 21 Stockholm, Sweden
Dr. A.O. Stuanes	Norwegian Forest Research Institute (NISK) The Agricultural University of Norway P.O.Box 61, 1432 Ås-NLH, Norway
Mrs. C. Sund	I. Krüger A/S Rymarksvej 108, 2100 Copenhagen Denmark
Mr. L. Vråle	Norwegian Institute for Water Research P.O.Box 333, Blindern, Oslo 3, Norway

Mr. S. Øberg — Microbiological Institute
The Agricultural University of Norway
P.O.Box 40, 1432 Ås-NLH, Norway

Assoc. Prof. H. Ødegaard — The Technical University of Norway
7034 Trondheim-NTH, Norway

INDEX

-0-